BASIC STATISTICS FOR SOCIAL WORKERS

REVISED EDITION

Robert A. Schneider

University Press of America,® Inc.
Lanham · Boulder · New York · Toronto · Plymouth, UK

Copyright © 2010 by
University Press of America,® Inc.
4501 Forbes Boulevard
Suite 200
Lanham, Maryland 20706
UPA Acquisitions Department (301) 459-3366

Estover Road
Plymouth PL6 7PY
United Kingdom

Library of Congress Control Number: 2009936463
ISBN: 978-0-7618-4932-2 (paperback : alk. paper)
eISBN: 978-0-7618-4933-9

The author wishes to thank Bradley Hjelmeland and Amanda Moore for editorial help.

For Tom and Terry

Contents

Preface to the Revised Edition

Through feedback from students the revised edition of Basic Statistics for Social Workers has been substantially improved from the first edition. Some of the earlier material has been deleted as redundant and some content has been expanded. In particular *Chapter Ten* on the *F* and *t-tests* has been expanded considerably to better explain tests of group independence for multiple groups. The chapters on bivariate and multiple regression have been more simplified to reflect the degree of difficulty that can be mastered by undergraduate students in social work. The chapter on single-system design has been clarified.

The pages are larger and the font size has been increased to 12 pt. Visually and conceptually the second edition is easier to follow. I use it as the required text in the statistics elective that I teach every semester and find no need to go beyond it.

Robert A. Schneider PhD, LISW
Department of Social Work
University of Northern Iowa
Cedar Falls, Iowa   50613
July 27th, 2008

For the Instructor

Basic Statistics for Social Workers, Revised Edition is a user-friendly text designed for teaching statistics to the undergraduate or graduate student of social work. The content is standard for introductory level statistics for social sciences. As much as possible, examples and problems are taken from typical field practice and agency-based or educational settings. Each chapter is relatively brief with more of an emphasis on understanding concepts in an ordinary way without a high degree of mathematical formulae. Experience has shown that social work students learn quantitative concepts better when the principles are attached to concrete examples rather than abstract applications. As such, this text should not be considered the equivalent of an introductory statistics course in a department of mathematics. The instructor may add as much mathematical material as he or she sees fit. Each chapter has problems at the end which are intended to reinforce the material.

Chapter One

Introduction and Variables

Statistics may be defined as *a branch of mathematics that involves the study of the frequency and distribution of numerical data*. We are exposed to statistics every day. "What percent of households have a working mother?" "What proportion of parolees are recidivists?" "Is there a correlation between sex and depression?" "What's the temperature outside?" "What is your GPA?" These are all questions that require answers in number form.

Reports or statements that simply describe numerical *data* (singular *datum*) are called *descriptive statistics* and are covered in the first five chapters. The other major branch of statistics is termed *inferential statistics*. In inferential statistics a researcher takes data from a sample and then generalizes to a larger group or population. Inferential statistics are covered in Chapters Six through Eleven.

Students who are often anxious about statistics and research methods often ask "But why do we have to learn this stuff. We're never going to use it!" I say "Not so". There are a number of valid reasons to learn some statistical methods.

First Reason to Learn Some Statistics

Social workers and other applied social sciences need to show that interventions and programs work. Measuring outcomes of a single subject or group intervention may help to do this. So the key word today is *outcomes*. Both social work practitioners and program managers may be required by law or funding organizations to demonstrate statistically that what they do with clients actually helps that is, there are improved outcomes. We are held to be more responsible to the taxpayers and funding sources. We are expected to show 'bang for the buck'. This is long overdue as well as inescapable in today's era of scarce resources.

Second Reason to Learn Some Statistics

We need to show with some confidence that in fact we really are helping clients. In part statistical methods can do this. This is similar to the first reason but for a different purpose. We've always just assumed that we were helping. In fact social work methods may not always help and may actually harm clients just as some medical interventions may harm patients, hopefully not too often. We have a responsibility to use the most effective methods possible when working with clients and we can't begin to know which methods work best unless we analyze practices or program outcomes statistically. This can help in our choice of methods.

Third Reason to Learn Some Statistics

As professionals we should be able to read and understand research results. In addition, we should be able to interpret research results for the average consumer who may not have a background in statisticsl. You may have noticed that newscasts, advocacy proponents and even your professors routinely give all kinds of statistical reports to audiences that knows little or nothing about the quality of the research reported. As future social science professionals you should be able to ask critical questions about the reported findings both for yourself and others who are uninformed.

Qualitative vs. Quantitative Data

The data used in research comes in two forms, qualitative and quantitative. *Qualitative* data adds dimensions beyond simple counting. Examples include case studies, sociological observation, anthropological observations, the content of speech or writing and the like. *Quantitative* data examines quantities or numerical data. Frequencies, percentages etc. are quantitative. Both qualitative and quantitative data are useful. The best research includes both when possible. Qualitative data is not considered scientific. This doesn't mean it's useless however. The scientific method includes the possibility of replication. Qualitative data is subjective and idiosyncratic for the most part. It cannot be replicated but can offer an understanding not found in numerical quantitative data and give us a subjective idea that leads to quantitative research. This text covers quantitative data only. There are research courses for qualitative studies usually as graduate research electives.

Variables and Attributes

Quantitative methods analyze the frequency and distribution of variables. Very simply put, a *variable* is something that 'varies', in type or degree. So for example, if you have male and female clients then SEX would be a variable. That is, SEX would vary depending on whether you had a male or female client. The variable SEX[1] could have one of two possible responses, or attributes. An *attribute* of a variable is one of the possible responses. Think of an attribute as a subset or category of a variable... Here's another one. If you have your temperature taken on Friday and its 98.2^0 and on Saturday it's 97.9^0 you would say that BODY TEMPERATURE varies. It is a variable. Let's look at some examples of variables whose attributes vary.

> Example One – Your clients, both men and women
> Variable = SEX
> Attribute = male or female
> Example Two – Religion of your clients
> Variable = RELIGION
> Attribute = Catholic, Protestant, Jewish, etc.
> Example Three – Number of children your clients have.
> Variable = NUMBER OF CHILDREN
> Attribute = 4 or 5 or 0[2] etc.

---

1) In the text variables will be capitalized throughout.
2) You usually don't hear of a score or count for continuous (number line) data referred to as an attribute, but it is nevertheless. Instead it's usually termed a *value* of *x*.

There are dozens and dozens of variables that we encounter in everyday life. But if all of your clients were female then SEX would not be a variable. It has only a single possibility. It is a *constant*. If you were to study the entering freshman class regarding high school courses then COURSES would be a variable but CLASS CATEGORY (freshmen) would be a constant since you are only studying freshmen class category does not vary. Chapter Two covers the types of numerical variables.

Summary

1. There are several reasons to learn statistics.
2. There are two types of data: qualitative and quantitative.
3. Quantitative data is necessary for the scientific method.
4. A datum can be a variable or a constant.

Chapter One - Practice Problems

1. If you are calculating the age of adolescent females in a community-based pregnancy prevention program are AGE and SEX variables? Explain.
2. What are typical attributes for the variable EYE COLOR?
3. Name four variables that might be of interest in a child welfare agency.
4. Name three variables from each of the following areas of research.
    a. Family functioning
    b. Inmate behavior in a minimal correction facility
    c. Depression
5. Suppose you are working with cancer patients. What are some qualitative phenomena that you might want to study?
6. Besides the reasons given in the chapter can you think of another reason to learn statistics?

Chapter Two
Levels of Measurement

Numerical data, the stuff of statistical analyses comes in different forms called *levels of measurement* (LOM). You can use the mnemonic device *NOIR*, French meaning 'dark' to remember the four levels of measurement:

> **N** nominal (categorical)
> **O** ordinal (categories in order)
> **I** interval (number line without a true zero)
> **R** ratio (number line with a true zero).

Statistical data is numerical data. Some calculations can be performed by hand with a calculator. But usually statistical data is analyzed with a software program of which there are many. One of the more popular packages is SPSS (Statistical Package for Social Science). SPSS and related programs can't analyze words. Therefore any attributes of a variable **must** be in numerical form. The exception is what's known as a *string* or text variable. Usually a string variable is a name or address etc. and is not analyzed numerically.

Nominal Level Data

Our first LOM (level of measurement) is *nominal* level data, also termed categorical. Think of it as a "bucket" of data where each bucket of data represents an attribute. Look at Image 2-1 below. Image 2-1 is a visual representation of the variable RELIGION. Remember that each attribute <u>must</u> have a number assigned to it. In this case the attributes or categories are numbered 1, 2, 3, and 4. I deliberately put the categories in non-sequential order to make a point. Since there is no natural hierarchy for religion it doesn't matter in which order you put the categories.

But we must assign some number to each attribute of RELIGION so that we can do calculations. The value you assign to an attribute for nominal level data is arbitrary however. If we code the attributes Catholic = 2 and Jewish = 1 that doesn't mean that Jewish is better than or higher or lower than Catholic. For nominal data you could just as easily code Jewish = 1, Others = 2 etc. and it makes no difference. One other thing – in nominal data categories are mutually exclusive. The requirement is that you can only be in one category at a time. You couldn't be coded as both Jewish and Catholic. There are no half-way points. Nominal data do not go on a real number line. And although we might have a range of numbers for the attributes the range of numbers is not a real number line as we would have for the variable AGE IN YEARS or HEIGHT IN INCHES. The attributes are just buckets of mutually exclusive data with no logical sense as to their order.

There is a case of the nominal variable called the *dichotomous nominal*. As the term dichotomous implies, it is a nominal level variable that has two attributes or categories (no more, no

Image 2.1 *Multicategorical Nominal Data*

less). Some examples are SEX (male-female), GRADUATED HIGH SCHOOL (yes-no) etc. You get the point.

Ordinal Level Data

Ordinal data is sort of a cross between buckets of data and data on a number line. It is data that is rank-ordered, that is, the data is put in a hierarchy. "Win, place, or show" in the Kentucky Derby is an example. Like number-line data there has to be an order from lowest to highest but you never know the actual distance between the first place and second place horse for example. All you know is one horse came in first, one second, and one third. Like nominal level data, the categories or attributes are mutually exclusive. "Oldest, middle, or youngest" is another example of ordinal data (see Image 2.2) that is, SIBLING POSITION. Yet another would be agreement to a statement where the possible responses are "strongly disagree, disagree, not sure, agree, strongly agree." The assumption for ordinal level variables is that there is a "hidden" number line for the attributes.

Image 2.2 *Ordinal Data*

Interval Level Data

Interval and as we shall see *ratio* data is simply put, data that goes on a number line. Some examples are AGE, HEIGHT, NUMBER OF CHILDREN, NUMBER OF CLIENTS etc. In this case the 'attributes' if you will, are simply numerals. They do not have categories tied to their numerical code as in nominal and ordinal level variables. On any number line the "hash marks" are spaced evenly apart. So the space between 1 and 2 is the same as the space between 77 and 78. This level of measurement does not have true natural zero. A carpenter's ruler (see Image 2.3) is an example of an interval level measurement tool. There is no zero on a carpenter's ruler. We don't usually measure a distance of zero inches. Height in feet is another. You can't be zero feet tall. It makes no sense does it?

Image 2.3 *Interval Level Data*

Ratio Level Data

The only difference between interval and ratio data is that interval data does not have a true zero, i.e., you can't be zero inches tall; and ratio data does have a true zero - you can have zero children. But for our purposes there is no need to distinguish between the two levels of measurement. Except in some advanced statistical methods you can use either interval or ratio the same way in formulae. It is common to use the term *interval-ratio level data*.

Summary

1. There are four levels of measurement.
2. Nominal and ordinal are variables with mutually exclusive attributes.
3. Ordinal data is simply categorical data with its attributes in order.
4. Interval-ratio data is data on a number line.
5. Ratio data has a true zero. Interval data does not.

Chapter Two -Practice Problems

Identify the following variables with the correct level of measurement.

- a. FREQUENCY OF HOSPITALIZATION
- b. NUMBER OF VISITS TO your AGENCY
- c. BEING A SUBJECT IN A STUDY that has a control and an experimental group
- d. OLYMPIC MEDALIST
- e. PROPORTION OF CLIENTS ON TANF
- f. Your GPA
- g. Your EYE COLOR.
- h. INTEREST ON A LOAN
- i. RATE OF DECAY OF CARBON-12 ISOTOPE
- j. Your BLOOD TYPE

Chapter Three
Data Representation

You should always visually inspect your data before conducting any statistical calculation. There can be data entry errors, missing data or data that don't meet the assumptions underlying a given calculation. This inspection can be done with various tables, graphs and charts, some of which you're familiar with. There are other times when you merely want to understand the simple distribution of a variable – perhaps the range or perhaps the most frequently occurring attribute or score, the mode. For this text we will cover frequency distributions, crosstabs, percentiles and histograms first.

Frequency Tables

A *frequency distribution* is one visual way to understand data and is one that you may have seen and which you should find fairly easy. Suppose you were to collect data on a number of subjects. In a frequency table all those having the same score, value or attribute for a variable are grouped together in the same row in a frequency table. The frequency table can display any level of measurement. There are two examples below. One for the dichotomous nominal variable SEX and one for the ratio variable NUMBER OF CREDITS ATTEMPTED.

Table 3 is a typical frequency distribution for the dichotomous nominal variable SEX. It has two categories. Notice that each attribute (male, female) of the variable SEX has the number of subjects or the frequency for a particular attribute. To the right of frequency is the *percent* of the total that the attribute represents. For example in Table 3.1, the 14 males are 31.1 % of the whole group of 45 subjects. On the far right is a *cumulative percent*, i.e., total up to that point, a running total that includes the row where the total is.

Frequency Distribution for the Variable SEX
Sex of cases (1=Male, 2=Female)

Value Label	Freq.	Percent	Cum. Percent
1	14	31.1	31.1
2	31	68.9	100.0
Total	45	100.0	100.0

Valid cases 45   **Missing cases** 0

Table 3.1 *Frequency Table for* SEX

Our second example is NUMBER OF CREDITS ATTEMPTED for a group of 55 students. Notice the value 12 in Table 3.2 below. This *12* is the number of credits attempted for each of eight students from the sample of 55. These eight students represent 14.5% of all 55 students and the cumulative total at this point is 20%. The cumulative frequency includes not only the eight students taking 12 credits but also all the other students taking less than 12 credits.[3]

Frequency Distribution for the Variable
NUMBER OF CREDITS ATTEMPTED

Value Label	Freq.	Percent	Cum. Percent
0	1	1.8	1.8
6	1	1.8	3.6
10	1	1.8	5.4
12	8	14.5	20.0
13	7	12.7	32.7
14	1	1.8	34.5
15	19	34.5	69.1
16	9	16.4	85.5
17	4	7.3	92.7
18	1	1.8	94.5
19	1	1.8	96.4
20	1	1.8	98.2
21	1	1.8	100.0
Total	55	100.0	100.0

Valid cases   55   Missing cases   0

Table 3.2 *Frequency Table for* NUMBER OF CREDITS ATTEMPTED

Sometimes you will see a frequency table with columns labeled 'percent' and 'valid percent." Or "cumulative percent." You want to consider only the valid percent since it does not include the missing data as you can see in Table 3.4, next page. Notice that the 7 missing values make up 7.4% of the total

Crosstabs

Crosstabs or crosstabulation (using a table) is another way to understand your data. Crosstab analysis is almost always limited to categorical data (nominal and ordinal). In a crosstab two categorical variables intersect in rows and columns. The example below (Table 4.2) is a reprint from an SPSS crosstabs of two variables SEX and EMPLOMENT

---

3) Due to rounding error the cumulative percent may not be 100. 0.

EMPLOYMENT CATEGORY

		Clerical	Custodial	Manager	Total
SEX	Female	206		10	216
	Male	157	27	74	258
Total		363	27	84	474

Table 3.3 *Crosstabs of* SEX * EMPLOYMENT CATEGORY

If you look where the SEX variable crosses the EMPLOYMENT CATEGORY variable with its three attributes 'clerical', 'custodial', 'manager' you can detect patterns such as: There are more female clerical employees than male; There are 27 male custodial employees and no females in this position (an empty cell is a zero), etc. If you simply look at the two variables alone all you will have is the total number of employees in each category.

The 'totals' down the far right column and across the bottom row are called the *marginals*. So you total down or across to get either the row marginal or column marginal. The grand total is in the bottom right cell.

AGE IN YEARS

		Freq.	Percent	Valid Percent	Cum. Percent
Value	18.00	7	7.4	8.0	8.0
	19.00	22	23.2	25.0	33.0
	20.00	19	20.0	21.6	54.5
	21.00	18	18.9	20.5	75.0
	22.00	12	12.6	13.6	88.6
	23.00	5	5.3	5.7	94.3
	24.00	2	2.1	2.3	96.6
	25.00	3	3.2	3.4	100.0
	Total	88	92.6	100.0	
Missing	System	7	7.4		
Total		95	100.0		

Table 3.4 *Frequency Table for* AGE IN YEARS

Histograms

Another way to visually inspect your data is the *histogram*, one of most common no doubt[4]. It's pretty much the same as a bar chart that you're already used to. That is, the frequency or percentage

---

4) There are other graphic representations of data such as the *stem-and-leaf* to name one. I chose to omit them here.

of subjects for each value on the **x** or horizontal axis is represented by a vertical bar. The frequency (count) or percentage is represented by the height on the **y** or vertical axis. The values or attributes are side-by-side in numerical order on the **x-axis**. Like the frequency table, you can use a histogram for any level of measurement as long as you have a bar for each attribute or score. There are two examples below.

Figure 3.1 below is a histogram for the GRADES for a class of students. Grades are grouped in 5 point intervals on the **x axis**[5]. For example, the value 85 (as a midpoint) represents all the grades in the interval bounded by the values 82.5 and 87.4. The left vertical axis (the **y axis**) tells you how many are in each five increment group on the horizontal x-axis. Twelve students earned between 77.5 and 82.4 where 80 is the midpoint. We cannot tell from the histogram what the scores were in each case, only the number of students within each 5-point interval and the midpoint or average of the five in each increment.

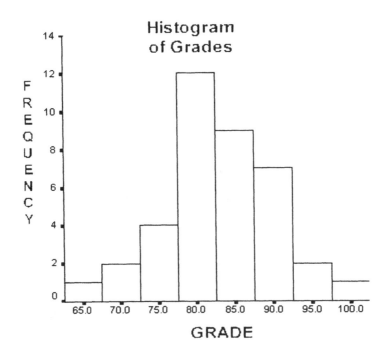

Figure 3.1 *Histogram for* GRADES

Figure 3.2 below is a histogram for an ordinal variable, SIBLING POSITION. As we said, a histogram as well as a frequency table can be used for *any* level of measurement. There are twenty-six first-born subjects in this group of 55.

---

5) An *interval* is a mini-range of scores, usually collapsing 5 or 10 possible values into a single *midpoint.*

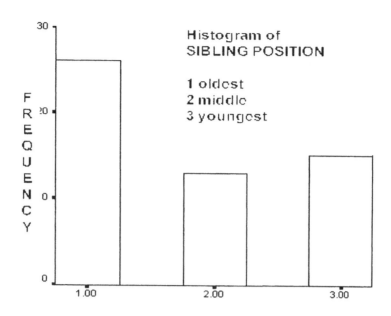

Figure 3.2 *Histogram for* SIBLING POSITION

Percentiles

As students who have taken various standardized tests, I'm sure you're already familiar with the idea of percentiles. A percentile is a value on a scale from 1-100 that indicates the percent of a distribution for a score that is equal to or above a particular value or raw score. So a percentile of 77 represents the score or value that exceeds or equals 77% of all the other scores. A percentile of 82 represents a raw score or value that is higher than or equal to 82% of all the other scores.

This is one way to standardize data by the way. That is, relatively speaking it doesn't matter how bad or how good all the scores are and it doesn't matter what the range is. Standardizing in this sense simply adjusts the values relative to all the other scores in the data. So there is no actual or raw score for a percentile.

You may take the range of the percentile and divide it into sections like grouped frequencies. Quartiles (25% increments - there are four), quintiles (20% increments – there are five) and deciles (10% increments – there are 10) are the most useful. Look at the Table 3.4 below. This is another version of the continuous variable NUMBER OF CREDITS ATTEMPTED that we saw earlier in the frequency distribution section but the range of scores is divided into quartiles, ¼'s. I deliberately chose the 25th, 50th, and 75th percentiles to illustrate the idea of a quartile. But I could have just as well shown the whole 100% as percentiles as above in Table 3.2. Choosing the 100% method is the only way to associate percentile with a particular score. The other methods give you scores for data that is grouped by quartiles deciles etc.

NUMBER OF CREDITS ATTEMPTED

Percentile	Value
25	13.0
50	15.0
75	16.0

Table 3.5 *Percentiles as Quartiles*

OK. So to interpret a score as a percentile, just take the score in the right column, 13 credits for example in Table 3.4 and move left to see which percentile is associated with it. It's that simple. 13 credits would put you at the 25th percentile of all students in the sample. Given our definition of percentiles the 25th percentile includes all those who are attempting 13 credits or less. A student who is attempting 15 credits would at the 50th percentile and 16 credits would be at the 75th percentile. Percentiles only make sense for *interval-ratio data*. It makes no sense to rank order eye color etc.

Summary

1. Always inspect your data visually before doing executing a statistical test.
2. Standard ways of graphically representing data are frequency tables, crosstabs, percentiles and histograms.
3. Converting a score to a percentile is a way of standardizing the scores or values.

Chapter Three – Practice Problems

1. (True or False) Any level of measurement can be visualized with a frequency table.
2. Why don't we have zero as a percentile?
3. (True or False) The 40th percentile is the point which exceeds 40% of all the scores.
4. For the following table what is the mode and what percent of scores does it represent? What is the cumulative percent for the value 13? How many subjects have a value of 13?

Value	Frequency	Percent
0	1	1.8
6	1	1.8
10	1	1.8
12	8	14.5
13	7	12.7
14	1	1.8
15	19	34.5
16	9	16.4
17	4	7.3
18	1	1.8
19	1	1.8
20	1	1.8
21	1	1.8

5. State two conclusions for each of the two tables, one for RELIGION and one for SEX.

RELIGION

		Freq.	Percent	Valid Percent	Cum. Percent
Valid	Catholic	32	33.7	33.7	33.7
	Protestant	28	29.5	29.5	63.2
	Jewish	3	3.2	3.2	66.3
	Other	32	33.7	33.7	100.0
	Total	95	100.0	100.0	

SEX

		Freq.	Percent	Valid Percent	Cum. Percent
Valid	male	26	27.4	27.4	27.4
	female	69	72.6	72.6	100.0
	Total	95	100.0	100.0	

6) From the histogram for GRADES in Figure 3.3 on page 4, approximately how many   students have a grade between 72.5 and 77.4? Using this estimate what is the approximate percentage of the whole that this frequency represents?

Chapter Four
Measures of Central Tendency

Now that we've seen a few ways of inspecting data, we can look further at ways that data can be organized. This may be accomplished with a few simple formulae[6]. We'll start by examining how the data (plural) group together around the middle of a distribution of scores or values for a given variable, i.e., the *measures of central tendency* for a variable. In the next chapter we'll examine how data are spread out, i.e., the *measures of dispersion*.

The standard representation of interval-ratio data is the *bell curve* (shaped like a bell). Nominal and ordinal data can also be viewed on the bell curve but with some limitations that we shall soon see. Look at the bell curve (also called the *normal curve*) in Figure 4.1. There are three measures of central tendency: the *mean*, the *median*, and the *mode*. Measures of central tendency are essentially what we usually think of as the average, or even the typical value. The average or typical scores fall at or near the middle point or 'hump' on the curve. Now this is important – while all three measures of central tendency can be viewed on the curve for an I-R variable, only the mode can be assigned a value for nominal or ordinal (categorical) data. This is because "buckets of data" are only categories of data. You can execute any of the four mathematical functions on categorical data since nominal and ordinal data are not understood as having an underlying scale or number line. So let's look at each measure of centrality one at a time and this puzzle will make more sense.

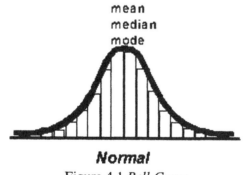

Normal

Figure 4.1 *Bell Curve*

Mean

The *mean* is what we usually think of as the average, the arithmetic average that is. You could get the mean or average age of all the students in your class or the average age of all your clients by adding up everybody's age and dividing by the number of people or subjects. There is a formula for this and it is the first formula you need to know.

---

6) Formulae and equations are actually shortcuts, a freeze-frame or some relationship between or among data symbols.

$$\overline{x} = \frac{\Sigma\,x}{n}$$

Formula 4.1 *Mean of a sample*

The symbol on the left is called *bar-x* $\overline{x}$ and is the symbol for the mean. Σ (sigma) means 'to add up whatever follows'. **n** means the number of subjects. **x** is a score or the value for a variable which in this case is a subject's AGE. So you add up all the ages (Σx) and divide by the number of subjects (**n**) to get the mean $\overline{x}$. Simple.

A sample is a group of subjects for whom attributes and their frequency are known. We have them; we know who they are; we can see them. Ideally this sample is representative of a larger population that we do not have. We don't actually have the population from which the sample was drawn but we can hypothesize or estimate the mean of a population μ (the Greek letter Mu) based on a sample, but with limited confidence, i.e., with some chance of error. More on this later. The formula is identical to the sample mean formula but μ replaces $\overline{x}$.

$$\mu = \frac{\Sigma\,x}{n}$$

Formula 4.2 *Mean of Population*

Median

This is easy too. The *median* is just the *middle score* for all the **x**'s. You've probably heard of the 'median family income'. This means that 50% of families have an income above the median, and 50% below. It's the midpoint in other words. When **N** is large (30 or more) you may simply identify the 50th percentile as the median or midpoint. But for a smaller N there are two ways to find the middle score.

The first way is to use a simple formula. Subtract the lowest score from the highest score, divide by two, then add the lowest score to get the median. This method should not be used with small samples <30.

$$\text{med} = \frac{(H + L)}{2}$$

Formula 4.3 *Median*

For smaller samples use the following method of simply picking the middle score. But there's a trick to this. You must put your scores in order first (*rank order*). For example, if you have seven scores:

1, 3, 1, 7, 4, 1, 6

you must first order them as:

1, 1, 1, 3, 4, 6, 7

before picking the middle score 3 as the median. This is the method for a small **n** with an odd number of subjects. But what if you have 6, 8, 1, 9, 10, 5, an even number of scores? If you order them:

1, 5, 6, 8, 9, 10

there is no middle score! In the case of an even number of values like this you take the average of the middle two scores no matter what they are. So the average of the two middle score 6 and 8 is 7. So 7 is the median.

It might seem odd that the median in the first example above is 3 or any number that is not really at the absolute center of the range of scores. Even though the median can be calculated in this way in practice there is little or no use for a median in small samples. Taking the mean as the 50th percentile is more the case for a larger number of scores such as median family income for the nation.

For any size sample statistical software will calculate the mean, median and mode (next). You don't need to do it by hand as a rule.

Mode

The *mode* or *modal score* is simply the value or score *that occurs the most*. By hand, all you need to do is put your data in a frequency distribution or histogram to detect this. In the case where there are two scores which both occur more frequently than other scores, then you can say that your data is *bi-modal* or *multimodal*. Either way is correct. In Table 4.1 below you can see that the most frequently occurring value or score is 15. There is a count or frequency of 19 for this value. So the mode is 15.

As mentioned above, categorical data only has a mode since you can't add up or divide unique categories such as SIBLING POSITION in Figure 4.2, next page. You can tell from this histogram that the most frequently occurring score or attribute is 'oldest'. 'Oldest' is the mode. There is no mean or median.

Value	Frequency
0	1
6	1
10	1
12	8
13	7
14	1
15	19
16	9
17	4
18	1
19	1
20	1
21	1

Table 4.1 *Mode*

Statistics and Parameters

You saw above that the symbol for the mean of a sample is $\bar{x}$ and the symbol for the mean of a population is μ (mu). Besides the mean we will be going through several statements involving symbols for both samples and populations. For example, a correlation between two variables in a sample is **r** while the estimated correlation between the same two variables in the population is ρ (rho). These measures of a sample are termed *statistics* while their equivalent values in the population, estimated from a sample, are termed *parameters*. So **r** is a statistic and ρ is a parameter.

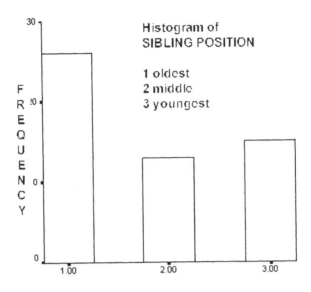

Figure 4.2 *Histogram for* SIBLING POSITION

Skewness

Sometimes data is not normally distributed. That is, the hump that's usually in the middle of the bell curve may be shifted to one side or the other. The term for this is *skewness* or skewed data. When the 'tail' stretches out to the left it's termed left skewed or negatively skewed since the direction of the tail determines the name for the skewness. If the tail goes out to the right it's called a positive skewness. See Figures 4.2 a, b, and c below.

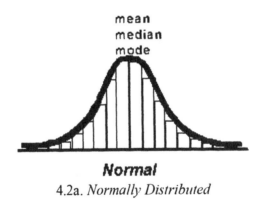

4.2a. *Normally Distributed*

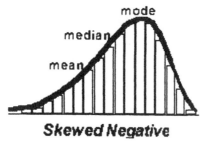

Skewed Negative

4.2b. *Negative or Left Skew*

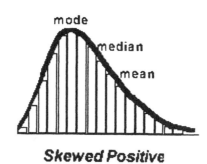

Skewed Positive

4.2c. *Positive or Right Slew*

Notice that the mean, median, and mode go 'up the slope' either from the right or from the left. Most data are in fact skewed a bit so the mean, median, and mode are rarely the exact same value. But it's a question of degree. When the skew is significant (and there is a way to tell this), then the variable has to be transformed more to normality. This is usually done algebraically or logarithmically. Transforming significantly skewed data is a little beyond this text but you need to be aware of the fact that in most cases significantly skewed data can't be properly analyzed until it is transformed first.

Summary

1. There are three measures of central tendency: the mean, median and mode.
2. All three measures may apply to interval-ratio data.
3. But only the *mode* applies to nominal and ordinal data.
4. Statistics such as "bar-x" apply to samples. Parameters such as μ apply to populations.
5. Skewness is the measure of how much the 'hump' of the bell curve is off-center or lopsided.

Chapter Four – Practice Problems

1. If the mean, median, and mode for number of children is in the following order 2.67, 2.00, 1.27, is the distribution of scores likely skewed? If so, in which direction?
2. If the mean, median, and mode for number of children is in the following order 2.67, 2.50, 2.27, is the distribution of scores likely skewed? If so, in which direction?
3. For a large number of scores the median is likely to be the ___ th percentile.
4. What can you say about a curve when the mean, median and mode are near each other?
5. Calculate the mean, median, and mode for the following lists:
     a) 166, 23, 10, 344, 87, 88, 102, 8, 68, 87, 35, 79.
     b) 7.233, 0.003, 12.099, 7.000, 7.000, 124.879, 0.000.

Chapter Five
Measures of Distribution

Just as scores will group around the middle as measures of central tendency, so will they have a spread on both sides of the middle. These measures are known as *measures of dispersion* or *measures of distribution*.

Range

The *range* is the easiest to understand. It's simply the *gap between the highest and lowest scores.* Once you put all the values or scores in order you just calculate the difference between the highest and lowest, or simply state the highest (*maximum*) and lowest (*minimum*) score. Look at Example 5.1 below. There are seven scores which you put in order first as when you calculate the median. For example given the scores:

$$1, 1, 1, 3, 4, 6, 7$$
$$\text{maximum} = 7$$
$$\text{minimum} = 1$$

$$\text{maximum} - \text{minimum} = \text{range}$$
$$7 - 1 = 6$$
$$\text{range} = 6$$

Since the range is the difference or gap between the maximum score 7 and the minimum score 1, the range is **6** (the difference between 7 and 1). Researchers often report only the minimum and maximum. Often this is more important to know than the actual range itself.

Standard Deviation and Variance

If you recall the normal curve (for interval-ratio data only) that we used in the last chapter, you know that there is a spread of scores under the curve, 50% on each side of the mean (Figure 5.1).

The curve will vary in shape depending on the range and the data for a particular variable. Sometimes the curve is flatter and spread out with a low hump (small *kurtosis*, as it's called); sometimes the hump is high and the tails of the curve don't spread out or disperse as far. It so happens that statisticians have agreed on a way to describe *given percentages of the area under the curve* as long as the curve is more or less normally distributed (i.e, minimal skewness, minimal *kurtosis*).

The term for a given area under the curve is the *standard deviation (***s** or **SD** or **std. dev***)*. The area is then reported in terms of the number of standard deviation units.

This is perhaps the single most important concept to understand in learning statistics. All statistical analyses using 'interval-ratio data', and there are many, use the *standard deviation* as the basic building block.

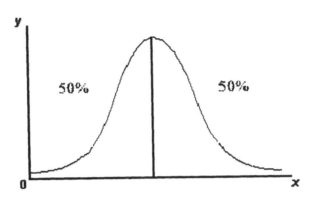

Figure 5.1 *Normal Curve*

But before we look at the formula for calculating the standard deviation, let's see what we mean by the areas under the curve (height x width). We'll start with whole units of **SD**, that is, either one, two or three standard deviations. Look at Figure 5.2 below. The dark area represents the area covered by one standard deviation above the mean and one standard deviation below the mean. You would correctly state that the area represents "plus or minus one standard deviation unit" or "plus or minus **1 SD**". The sloping tails of the curve never reach the x-axis. They continue + and – to infinity. The term for this normal curve then is *asymptotic*. This in theory a score could be any number of standard deviations from the mean all the way to infinity. In practice we usually exclude value greater than 2 standard deviations from the mean in either direction. Figure 5.3 is a graphic representing plus or minus 1, 2 and 3 standard deviation units. Figure 5.4 shows sections of the curve below between **–2SD** and **+3SD** units.

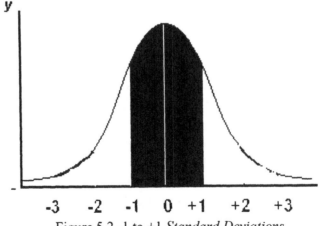

Figure 5.2 -1 to +1 *Standard Deviations*

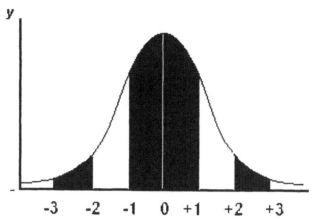

Figure 5.3. -3 to +3 *Standard Deviations*

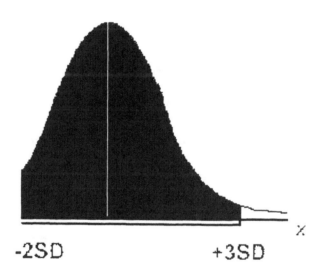

Figure 5.4. - *2 to +3 Standard Deviations*

Statisticians have determined given percentages under given specific standard deviation areas regardless of the variable. You need to memorize the following:

1. The area between + **1 SD** and – **1 SD** is 68.26 % (round to 68%)
2. The area between + **2 SD** and – **2 SD** is 95% (actually 95.44%)

3. The area between + **1.96 SD** and **-1.96 SD** is exactly 95%
4. The area between + **3 SD** and **-3 SD** is 99.74% (approx. 99%)
5. Of course, 50% of the total area is on each side of the mean

Now let's look at the formula for calculating first the variance, then the standard deviation, Equation 5.1. You already know all the elements. Before calculating the standard deviation itself you first calculate what's called *the variance* (S^2, Equation 5.1).

To calculate the standard deviation you first must calculate its square (*variance).*

$$s^2 = \frac{\sum(x - \bar{x})^2}{n-1} \qquad \text{or} \qquad s^2 = \frac{\sum(x - \bar{x})(x - \bar{x})}{n-1}$$

Equation 5.1 *Variance*

These are the steps:

1) After you calculate the mean, you subtract it from each score in your data set. This is called the deviation score. Remember this term, *deviation score*.

$$(x - \bar{x})$$

Equation 5.2 Deviation Score

2) Now square each of the deviation scores, i.e., the *deviation score squared.*

3) Then you add up all of the deviation scores squared. The result is the deviation *scores squared* or the *sum of squares* (**SS).**

$$\sum(x - \bar{x})^2 \qquad \text{or} \qquad ss = \sum (x - \bar{x})(x - \bar{x})$$

Equation 5.3 Sum of Squares (Deviation Score Squared)

4) Then you divide by the number of subjects minus one **(n-1)**[7] to get the variance (**s²)**). It looks more complex than it is until you break it down.

$$s^2 = \frac{\sum(x-\bar{x})^2}{n-1}$$

Equation 5.4 *Variance*

---

7) Use (n-1) for small sample, i.e., < 30. For larger samples just use 'n'.

Now that we've calculated the variance all we have to do is take its square root to get the standard deviation. For a sample with **n** subjects:

$$s = \sqrt{\frac{\sum (x - \bar{x})^2}{n-1}}$$

Equation 5.5 *Standard Deviation Formula for a Sample*

Let's do one! Suppose that you had clients' ages 10, 12, 12, 15, 18, 20. What is the standard deviation for these data?

$$s = \sqrt{\frac{\sum (x - \bar{x})^2}{n-1}}$$

- Calculate the sum: $10 + 12 + 12 + 15 + 18 + 20 = 87$

- Divide by **n = 6** to get a mean of **14.5**

- Subtract 14.5 i.e., the mean, from each score. This is the *deviation score*.
  1. $10 - 14.5 = -4.5$
  2. $12 - 14.5 = -2.5$
  3. $12 - 14.5 = -2.5$
  4. $15 - 14.5 = .5$
  5. $18 - 14.5 = 3.5$
  6. $20 - 14.5 = 5.5$

- Then square each deviation score to get all positive numbers.
  1. $-4.5 \times -4.5 = 20.25$
  2. $-2.5 \times -2.5 = 6.25$
  3. $-2.5 \times -2.5 = 6.25$
  4. $.5 \times .5 = .25$
  5. $3.5 \times 3.5 = 12.25$
  6. $5.5 \times 5.5 = 30.25$

- Add them up to get the *deviation scores squared* or the *sum of squares* (**SS**) which is **75.50**.

$$\sum (x - \bar{x})^2 = 75.50$$

- Divide by **n-1** which in this case is **5** (6 subjects -1) to get **15.10**. This is the *variance* for our scores. So $s^2 = 15.10$.

- But we need to get the standard deviation so we just take the square route of the variance which gives us a *standard deviation* of 3.89 or SD = 3.89, or **s = 3.89**.

$$S = \sqrt{\frac{\sum (x - \bar{x})^2}{n-1}}$$

so,

$$S = \sqrt{\frac{75.50}{5}} = 3.89$$

I only used six scores to make it easy. Actually you could never get a meaningful **SD** from such a small data set. The very minimum number of scores or values required is 10. When you have an **n** larger that 30, then you can drop the **(n-1)** term and just use **n** as the divisor.

The variance is difficult to explain. Yes, it is the **SD** squared, but cannot be shown on a number line or a 2D or 3D graphic such as a curve. It is a squared value of a SD area under the curve and can't be visualized. No concern now. Just know that the variance is **s²** and not **s**. Finally we mentioned parameters last chapter. The equivalent standard deviation in a sample is represented by σ (lowercase sigma) and uses μ (mu) and the unknown **n** and unknown **x**. This is hypothetical and cannot be easily calculated but it's necessary for inferential statistics as we shall see later.

$$\sigma = \sqrt{\frac{\sum (x - \mu)^2}{n}}$$

Equation 5.6 *Population Standard Deviation*

Z-Scores

As mentioned, statistical tests using interval-ratio numberline data rely on the SD as the basic building block of parametric[8] ("parallel to the meter", that is) statistics. Also as mentioned, the **SD** is used simply as a way to define given areas under the curve. We saw for example the **SD** in terms of whole number, 1,2,3 **SD's** etc. This will become clearer now as we cover the topic of **z-scores**.

The **z-score** is the number of standard deviations units from the mean for any score on the **x-axis** (domain). So if you have a value that is exactly two standard deviations above the mean, then that same point would have a **z-score** of exactly **+2**. It's not too hard to understand. By the way, converting any score to a **z-score** is the typical way to standardize scores into a common measure, this time the **z-score** rather than a percentile. The reason we term this a standardization method is that, like the percentile, **z-scores** are converted to a common language or *standard* distribution regardless of the variable. The mean for **Z** is zero (i.e., zero **SD's** from itself)

To calculate the z-score for any value of x all you need is the standard deviation and the mean that you've already learned to calculate. The formula for the Z-score for a sample is:

---

8) Analyses that use I-R data are termed *parametric statistics*. Those that use categorical data are termed *non-parametric* statistics. This is not to be confused with *parameters* which are measures in the population.

$$Z = \frac{X - \overline{X}}{s}$$

Equation 5.7 *Z-Score*

So for the same data earlier suppose that you're asked to calculate the **z-score** for the value **x = 12**. You already know the mean and the standard deviation, **14.5** and **6.97** respectively. All you need to do is plug the numbers in to the formula for **z**:

Mean = 14.5
SD = 7.32
x = 12
Z = ?

$$Z = \frac{X - \overline{X}}{s}$$

$$Z = \frac{12 - 14.5}{7.32}$$

$$Z = -0.34$$

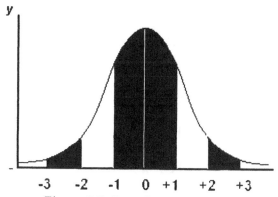

Figure 5.5. *Distribution of Z-scores*

Remember that a **z-score** *is* just *the number of standard deviation units that any score is away from the mean*. In the above problem you could correctly state that "the value **12** has a **z-score** of **-.34**" or "the value **12** is **.34 SD** units below the mean." Again the mean of any continuous variable always has a **z-score** of zero no matter what the actual value of the variable mean actually is (10 feet, 7.4 clients, 2.43 children etc.).

Area Under the Curve

As you may recall, a standard deviation unit represents a given area under the curve (68%, 95% etc.). So a **z** of +1 is at the point where the **SD** is also +1 and a **z** of +.56 is at the point where the **SD** is also .56. Since we know the specific areas for + or – 1, 2, and 3 standard deviations and we know how to convert any score to a **z-score**, we can then use the **z-score** to account for any part or fraction of the total area under the curve. All we have to do is calculate the **z-score** for any value and look it up in a table. *Table A* in the back of the text has a list of areas under the curve that can be calculated for a z-score. Only positive values for z are reported. It is understood that these can be
interpreted as negative scores when the area in question is below the mean. Let's use the example above. That is, the **z-score** for a value of 12 is -.34. One might ask what percent of the scores are between the mean of 14.5 and a score of 12 (i.e., a **z-score** of -.34). Below is a section of *Table A* that would be applicable. The relevant cells are in bold. On the far left column is the z-score of **.34**. In our case we know that this also represents -.34 since negative values are not given. You have to know that there is not a second table for negative numbers. The areas are in terms of percents. So 0.1331% of all the scores are between 14.5 and 12; that is, between the mean (14.5) and our calculated z-score of -.34. i.e., when **x = 12**.

Similarly beyond a z of -.34 out to the tail on the left is another 0.3669 % of all the scores. So by using this table we can find the area between the mean and any value of **x** by converting **x** to a z-score with the **z** formula.

z-score	Area From Mean	Area Beyond Z	z-score	Area From Mean	Area Beyond Z
0.32	0.1255	0.3745	0.67	0.2486	0.2514
0.33	0.1293	0.3707	0.68	0.2517	0.2483
0.34	**0.1331**	**0.3669**	0.69	0.2549	0.2451

Table 5.1 *Partial Table A*

To summarize from the table, for **x = 12** when **x = 14.5**
1. **z = -.34**
2. **0.1331** or **13%** is the area of under the curve from the mean (**14.5**)**12** in the negative direction.
3. **0.3669** or **37%** is the area of under the curve beyond **x = 12 (z = -.34)**

Standard Error of the Mean

There is another measure of dispersion that is often reported in addition to the range, variance, and standard deviation. Like the standard deviation it's a type of standardized score but it applies to the possible region or 'mini-range' for the *true* mean of the population. It's called the *standard error of the mean* (**SEM**).

First, understand that when you calculate the mean for a sample, the value of that mean is for that particular group of subjects only. So if you were to sample thirty clients from a roster of one-hundred and calculate the mean age for those thirty, would that mean be exactly the same as the mean for a second sample of thirty subjects from the same original one-hundred? Probably not. The true mean of the one-hundred subjects (i.e., the *population* or *universe*) will only be approximated by your sample of thirty (see Figure 6.1). The mean of a population is *mu* (μ). The mean of a sample is of course $\bar{x}$.

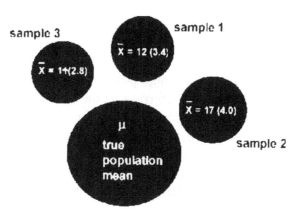

Figure 5.6 *Standard Error Graphic Model*

So why not just use the ages for all one-hundred subjects? Well if all you have is a hundred then you could. But suppose you have several thousand in your population or you're not even sure how many are in the population from which you drew your sample, or you only have enough money and time to sample forty or fifty or a hundred from the thousands in the population. Then as above, the mean for your sample would only approximate the true mean in the population of several thousand. In addition, even when you very carefully sample with each person having an equal chance of being selected from the population, you still only have a sample and cannot calculate the true mean of the population. Well there's a way around this, the *standard error of the mean*. The standard error of the mean is based on the *central limit theorem*. The *central limit theorem* It so happens that as you increase the sample size any variable will tend to become normally distributed, even if the original sample isn't, i.e., the *central limit theorem*. In other words the clump of data in the middle of the bell curve is typical for most any naturally occurring variable as sample size increases. The formula for the SEM is next. For a sample simply divide **s** by the square root of **n**. The quotient is added and subtracted from the mean. This will yield the mini-range where the true mean is estimated to be. This brings us to another point concerning the mean.

The mean as you know is just the arithmetic average. It does not need to be an actual value of a score **x**. It's a hollow shell, an empty suit. The problem is that in parametric statistics virtually all tests use the mean as the starting point. If you base conclusions on the mean or its derivative formulae don't bet the farm that it's true. It's just an approximation of the average. And not only that, it isn't even a true value for any subject except by accident.

Typically you want to extend that SEM to a larger population by inference. In that case the SEM for a population is appropriate. The problem however is that we don't usually have σ so we use the sample SEM as an approximation.

$$s_{\bar{x}} = \sqrt{\frac{s}{n}}$$

Equation 5.8 *Sample Standard Error of the Mean*

$$\sigma_x = \sqrt{\dfrac{\sigma}{n}}$$

Equation 5.9 *Population Standard Error of the Mean*

You've seen the **SEM** used more times than you might think. When political polsters tell you that in a two-way race candidate Jones is five percentage points ahead of candidate Smith and the "margin of error is plus or minus three points" they're referring to the standard error of the mean. In the case of your clients' ages for example, if you calculate the SEM to be plus or minus 3.2 years for a sample mean of 25 years and you add and subtract 3.2 years from the mean of the sample, then according to the SEM, the true mean for the population from which you drew your sample is somewhere between 21.8 years and 28.2 years. There's a very important feature of the SEM, the SD and other related components of statistical formulae. The larger the n the smaller the SEM of SD.

If you want to be more precise about the SEM or SD, just increase your sample size.

Anything divided by 100 will be smaller than the same thing divided by 10. Try it.

The SEM can be visualized on the bell curve with one distinction. In this case the curve is not a distribution of scores or **x's,** it's a distribution of all possible means. Look again at figure 6.1 above. The smaller circles represent several samples from the same larger population. For each sample, a mean was calculated yet each mean is somewhat different from the others. The *SEM is actually the result of averaging all possible means* (i.e., to infinity).

Now notice the formula for the SEM. You already know all of its components. Naturally, the hump in the middle is the area where some value is more likely to occur because the area is greater (Figure 6.2). The SEM is around the middle and not out in the tails. We say with some confidence that the true mean of the population is somewhere near the mean of the sample. Statisticians are particular about the area under the curve and they agree that the true mean of the population μ is near the sample mean $\overline{x}$, but most likely not identical to it.

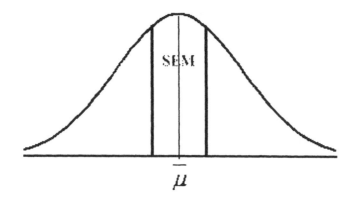

Figure 5.7 SEM *as Estimate of Population Mean*

Summary

1. Measures of dispersion are the *range, variance, standard deviation* and the *standard error or the mean*.

2. The *range* is the interval between the highest and lowest score and is the least useful. The *range* may be useful when the distribution of scores is highly skewed and the standard deviation meaningless.

3. The *variance* is the standard deviation squared. It is only reported in positive values since it is a squared number.

4. The *standard deviation* is the most useful of the measures. Since it is the square root of the variance it is reported in + or − terms.

5. The *z-score* is a transformed standard deviation. It equals the number of standard deviation units that some value **x** is away from the mean.

6. The *standard error of the mean* which is the + or − in polling error is sometimes reported in place of the standard deviation and is based in part on the standard deviation. It estimates a range where the true mean of a population falls assuming correct probability sampling.

Chapter Five – Practice Problems

1. Calculate the variance, standard deviation, range and SEM for the following two groups of IQ scores.
    101, 120, 98, 60, 146, 151, 100, 70, 120, 115, 112
    55, 134, 70, 110, 128, 90, 132, 120, 111, 107, 80, 87, 90

2. For the same two groups of scores find Z when IQ equals 110. When IQ equals 87. What is the area between the mean and each of these values of x?

3. What percent of the curve is between +1 and –2 standard deviation units?

4. What percent of the curve is between -1 and –2 standard deviation units?

5. What percent of the curve is below –2 standard deviation units?

Chapter Six
Inference

Sampling

See *Table A*, the *Table of Random Numbers* from the appendix. It is reproduced in part below. The numbers in the table were computer generated, i.e., purely randomly - absolutely no pattern. They are in groups of five so you won't go blind reading them. If you need a sample representative of a population the best way to do this is by mathematical or statistical sampling, i.e. using random numbers. You simply move across or down to pick however many numbers you need for you sample. It doesn't matter which way you go since they are random in all directions. So for example if you needed 10 subjects out of a population of 90 each of the 90 is assigned a unique number from 00 to 89 or 01 to 90. Then the first ten numbers that you come to in this range would comprise your sample. Using 01 to 99 and going down the left two columns (since 00 – 99 is two numerals wide) we get at random the numbers 39, 14, 30, 64, 42, 80, 05, 65 and continuing at the top for the next two-place numbers 63 and 59. If you needed a second round of 10 subjects you would start with the *73*. The rule for probability sampling is that each subject in the population has an equal chance of being picked. There is absolutely no pattern conscious or unconscious in picking the 10 numbers. In this way you can avoid any bias in your sampling.

Partial Table A
Table of Random Numbers

39634	62349	74088	65564	16379	19713	39153	69459	17986	24537
14595	35050	40469	27478	44526	67331	93365	54526	22356	93208
30734	71571	83722	79712	25775	65178	07763	82928	31131	30196
64628	89126	91254	24090	25752	03091	39411	73146	06089	15630
42831	95113	43511	42082	15140	34733	68076	18292	69486	80468
80583	70361	41047	26792	78466	03395	17635	09697	82447	31405
00209	90404	99457	72570	42194	49043	24330	14939	09865	45906
05409	20830	01911	60767	55248	79253	12317	84120	77772	50103
95836	22530	91785	80210	34361	52228	33869	94332	83868	61672
65358	70469	87149	89509	72176	18103	55169	79954	72002	20582

Table 6.1 *Partial table A, random numbers*

Mathematical or probability sampling (when the population is known) is the best way to select subjects who are representative of a population. Throwing darts at a list of numbers while blindfolded will produced biased results. It may not seem like it either consciously or unconsciously these unscientific methods of sampling can be subjective. The sure fire way to avoid bias is to use a table of random numbers or let the statistical software generate the random numbers or random phone numbers or random.........whatever as long as each eligible unit has a unique number.

There are further methods of random selection for different situations. You might have a lopsided population, 100 males and 700 females for example. In this case you use a *weighted sample* or *proportional sample* and sample seven times as many males assuming you needed a 50-50 sample based on SEX.   Sometimes researchers will group people geographically. For example if you needed 500 subjects from Iowa it might be better to use *cluster sampling* by clustering the population by the 99 counties in Iowa and then sampling proportionately from each county. Other related methods are *purposive sampling* and *stratified sampling*. Use of these methods depends on the purpose of the study and the nature or availability of the sampling frame.

In the real world of applied research such as social work research we usually do not have the luxury of true random sampling. Sometimes the population is too small, or unreachable. In research on service delivery it may be unethical to sample and then assign subjects to two groups, one which does not receive the services. In these cases we use what's called a *convenience sample* or an *ad hoc sample* or use the phrase *sampled as available*. This is not as desirable as statistical sampling and must be addressed in the writing-up of the results.

Probability and P-values

The earlier section on the standard error of the mean serves as a bridge between simply describing a sample (*statistics* for a sample) and relating it to the population (*parameters* for a population). The idea of inferential statistics, i.e., inferring from a sample that something is true in a population is really based on the probability that something is true or false in the population Pay attention to the probability that something is false. It will come in handy later. For our purposes now we'll use what's termed the *Z-test*. Don't worry about the details. Just know that the Z-test compares $\bar{x}$ to μ (the mean for a sample to the mean for the population) for a given known sample vs. its hypothetical population. We will hypothesize that $\bar{x}$ and μ are sufficiently different from each other to justify some statement about the difference. Here's how it works. First look at the four curves below. $\bar{x}$, μ and error, then for illustrative purposes all three on the same shared curve at the same time.

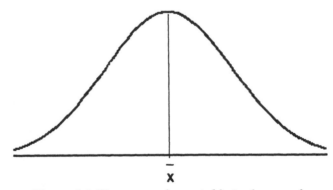

$$\bar{x}$$

Figure 6.1 *The mean of a variable in the sample*

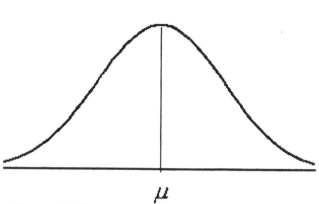

Figure 6.2 *The mean of a variable in the population*

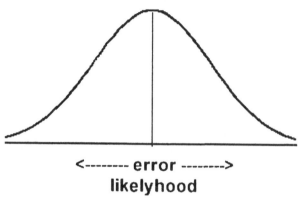

Figure 6.3 *The error distribution*

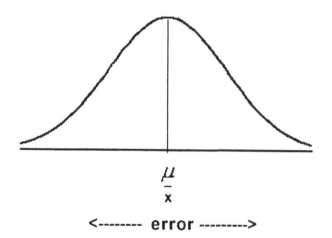

Figures 6.4. *Distribution of* **x** *, μ and Error*

So what we are doing is identifying (by an equation) the likelihood that **x** and μ are different to a degree that is not likely to happen by chance. And they have to be more than a little different. They have to be way different. In fact they have to be different enough that there is at the most only a 5%

chance that when we state that there is a difference (remember the probability that something could be false), that we are wrong. A 5% chance of being incorrect (the probability of making a false or erroneous statement) about the statement of difference is a risk that we are willing to take in social science. If we find out later that we were wrong in taking a 5% chance which we do in good faith, then we have made a Type I error. More on that in a bit. But if for a moment we put a 5% chance on the probability curve, we usually split 5% between the two tails of the asymptotic curve. Sometimes we may put the whole 5% in one tail but the two-tail split is the most common by far. This leaves 95% inside. For now just take this on faith.

Next it's necessary to understand what's known as the *null hypothesis*. The *null* or null hypothesis is the reverse of what the experimenter actually believes or suspects to be true *in the population;* a statement of something **not true** that is. It is deliberately put forward in order to be contradicted if possible (pr shown to likely be false). It's a "straw man". Since competent scientists are loathe to make statements of truth based on numerical data they thereby take the conservative perspective that we **must assume that nothing is going on** (hence *null*) rather than interpret results as if we actually knew what was true in a population that we don't have. In other words we hypothesize that the difference between the mean of the sample and the true mean in the population is negligible.

Since we don't have the population how can we say that something is true in the population based only on sample data? We can't! We must take the conservative view that nothing is going on and then eventually reject that in our z-test. Then we stop right there! We're done. That's it. This implies that something is indeed going on but doesn't really say it. We reject or do not reject the null (the probability that something is false). We **don't accept or state anything positive or even negative**. Simply reject or don't reject the null. That's how it works.

So now that you're thoroughly confused let's get back to our example of AGE and go through it step-by-step. This will make it more understandable. We'll assume that our sample is one of convenience. And we don't know if our sample mean is on the average higher or lower than the population mean. Follow the steps:

Step 1. Decide the tolerable probability of incorrectly rejecting the null. We don' have a clue as to whether our sample mean for AGE is higher or lower than the population mean for AGE. So we guess. That is we stipulate that a 5% chance of being wrong is all we'll tolerate. Since we don't know the direction of any difference between the ample mean and the population mean we use a 2-tailed test with 2.5% or .025 in each tail. This is called the 95% confidence interval. The 5% or in decimal terms .05 is known as *alpha.* So alpha is set to .05 by convention.

Step 2. Set up your *alternate* hypothesis, the opposite of the null but what you expect to possibly be true if you were to reject the null. For comparing a sample mean to a population mean for a 2-tailed   test the alternate is:

$$\mathbf{H_0}\text{: } \bar{x} = \mu$$

Equation 6.1, *Null Hypothesis*

In plain English this reads "There is no difference between the two means". Null, nada.

Step 3. Set up your *null* hypothesis, what you will reject or not reject if and when you detect acceptable statistical significance (safe probability). You have already set your statistical significance with alpha. Assume that we know both μ and $\bar{x}$. For comparing a sample mean to a population mean the *null* (no difference) is:

$$\mathbf{H}_1: \bar{x} \neq \mu$$

Equation 6.2, *Alternate Hypothesis*

If they are equal then there is no difference.

Step 4. Run the statistical test with its associated p-value (statistical significance). In this case, assuming we have some value for the population mean from previous record, we use what's called the Z- test. When we don't actually know what the population mean is we use the related *single sample t-test* instead. Forget the details of this for now. We'll come back to it later.

Step 5. Depending on which statistical test you perform software such as SPSS will give you the value of **z** or t (called the *test statistic*) as well as the **sig.** (*statistical significance* or *p-value*). As an example we might get a **z-value** of **3.78** and a **p-value** of **.047** when we compare μ with $\bar{x}$ with known population values.

Step 6. Reject or do not reject the null with the chance of being wrong, in this case as .047 or Probability of being wrong. (.047 is less than .05). You reject the null when your p-value is equal to or less than your beginning alpha or .05. 4.7% is less than 5.0% so you reject the null. To repeat, this says that "There is a 4.7% probability that when I reject the statement that the two means are different (null) that I will be wrong".

I'll repeat this over and over but it's **critical** - We use the null since we can't really say that the difference between the sample mean and population mean (and there will always be one) is significant. Not just different but significantly different from each other. What's more we can't even reject the null with one hundred percent certainty! There's always some chance that when we reject the statement of *no difference* and thereby imply or infer that there is something going on, i.e., a *true difference*, we might be wrong. If in fact we reject the null incorrectly, we've made what's called a *Type I Error*.

Recall that a 5% chance of something happening is assigned to the tails at 2.5% each, when alpha is set at .05 that is. If you set alpha at .10 then the tails would each cover 5% of the area under the curve. The calculated value of, in this case, **z** falls somewhere on the curve with an associated probability of making a type one error. The rejection region for **z** will vary depending on the number of subjects you have but in any case that **z** will have some probability no matter what. (See *Figure 6.5* next page).

Type I Error

The *null* or test hypothesis is symbolized by $\mathbf{H_0}$: It's opposite called the *alternate* hypothesis is symbolized by $\mathbf{H_1}$: The null hypothesis is the important one. The alternate hypothesis is only for completing the logic. By "completing the logic" we mean that all possible equalities must be covered. So if the null sates those two values are equal then the alternate must cover all other possibilities. In this case they are greater than or less than, in other words not equal. Be careful not to confuse not equal with null. *Null* usually mean equal or in other words *no difference*, hence null.

The probability of making a Type I Error is symbolized by *p*, and is reported in SPSS as *sig.* (meaning statistical significance). Graphically the p-value can be thought of in terms of likelihood of error and goes under the bell curve just like any other continuous variable. And the curve is asymptotic. When we mention a p-value we are really giving the probability or likelihood of making a Type I error. This can be any value between + and − infinity. A Type I error is incorrectly rejecting the null, or making a null statement that turns out later to be false. You have no control over this. It's just a natural probability. A roll of the dice if you will.

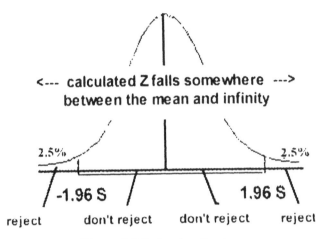

Figure 6.5 *Rejection Regions*

Type II Error

A *Type II error* is one over which you do have some control. A Type II error happens by not rejecting the null when you should have. As an example you might have a case where there really is a significant difference between the two means but you did not detect it when you should have. In other words the p-value didn't appear to be mathematically significant but was in fact just that. Why don't you see that? The most common reason is that you didn't have a large enough sample. If only you had more subjects you would have had a significant p-value. P-values and test statistics are sensitive to sample size. So if miss rejecting the null we have made a Type II error.

Confidence Intervals

Understand that for inferential statistics the area under the curve is not just a distribution of some variable, but rather the probability of making a Type I error. We call the region outside of a certain pre-determined area of the curve *alpha*, the *rejection region* (the region where you will reject the null that is) and we call the exact limits for that region the confidence intervals (C.I for short). Typically we use a 95% C.I., or a 99% C.I. (remember the area under the curve for standard deviations) etc. Look at Figure 6.5 for a graphic example of the 95% C.I. Actually 95% of the curve falls between ± 1.96 S. The area outside of the ± 1.96 S is called the rejection region. The whole area under the curve out to infinity represents all the likelihood or probability of incorrectly rejecting the null, or the probabilities of making a Type I error (zero to one-hundred percent).

Now the question becomes "OK, but what value is actually placed in or for that matter not in the rejection region (i.e., the area in the two little tails that totals 5% of the total are)?" The answer is that it depends on which inferential test you're doing. There are a number of inferential tests that yield a test statistic that you put on your probability curve to see if it falls in the rejection region or not. You've already learned about the mean. You might want to put the mean of your sample on the curve, infer that this mean is also true in the population (H_1:), see where if falls. To do this though you at least have to have a pre-existing estimated mean for the population first, μ_1. The next step is to place the mean for your sample $\overline{x}_1$ on the same curve. Then identify the point where the mean of the sample lands and decide by virtue of the area under the curve whether you want to reject the null or not. In this case we don't care if the mean of the sample is significantly above or below the population mean (i.e., a non-directional hypothesis), only that it's significantly different *in one direction or the other.*[9]

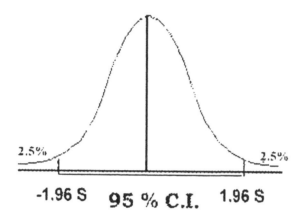

Figure 6.6 *95% Confidence Interval*

Let's use IQ for an example. Suppose that you have a sample of very bright students like yourself and you want to see if their mean IQ is significantly different from the general population. Of course it will be somewhat different since you just never have an exact match for the population. But will the difference be statistically significant is the question? You know that the mean IQ for the general public is 100. You calculate the mean and standard deviation for your sample to be 137 (SD,15). The C.I. for social science is usually –1.96 S to +1.96 S as we have stated. So you simply figure out where the 95% C.I. (p = .05) is for the population and see if your sample mean falls out in the tails (remember the two tails together equal 5%).

The next chapter starts the actual inferential tests that social work researchers commonly use. They are divided into two types. There are tests of group independence such as comparing the means for two groups, and there measures of association or correlations. We will start the later first.

Summary
1. Probability sampling is when each element has a known chance of being selected. Statistical sampling in general means that each element has an equal chance of being picked. Other related methods of statistical sampling are cluster, purposive and stratified sampling.

---

[9]In social science and social work research we don't typically reject the null unless the output in question falls outside the 95% C.I. In other words only if the particular outcome in question is way out in one of the two tails beyond –1.96 S or +1.96 S for a non-directional hypothesis will we reject the null with a 5% chance of making a Type I Error of course. Some disciplines use the 99% CI.

2. Statistical or probability samples can be obtained from a table of random numbers or can be computer generated.
3. *Ad hoc* or *convenience sampling* is used when you do not have the time or budget or it is unfeasible to sample statistically.
4. A p-value is the probability of incorrectly rejecting the null, i.e., a Type I error.
5. Alpha is the threshold value determined before data analysis that yields the likelihood for a Type I error. The obtained p-value then must be no larger than alpha.
6. Hypothesis testing is rejecting or not rejecting the null hypothesis of "no significant statistical relationship."
7. A C.I. is the range bounded by alpha on one or both tails. It is called a confidence interval because if your test statistic falls in that area then you are confident that there is nothing happening statistically. Therefore you don't reject the null. You reject the null when the test statistic falls outside of the C.I.

Chapter Six – Practice Problems

1. What is alpha and how do you determine it?
2. What is a p-value and how is it determined?
3. What percent of the curve contains the rejections regions in social work?
4. What is that percentage in terms of standard deviation units?
5. What does + or – regarding the rejection region mean?
6. Using Table A what are the first ten unique numbers picked when you want to sample 300 subjects?

Chapter Seven
Simple Correlations

Pearson Product-Moment Correlation

One class of inferential tests are named *measures of association* or *correlations*. It means as one variable changes so does a second variable. For example "The farther I run the more tired I get". For a correlation between two continuous variables, the test is called a *Pearson Correlation* (formally *Pearson Product-Moment Correlation*) and is symbolized by **r** which is the test statistic. For a correlation between two ordinal variables the test is called a *Spearman Correlation* and is symbolized by r_{sp}. And while it is little used, a correlation between a continuous variable and a dichotomous nominal is called a *Point Bi-Serial Correlation* and is symbolized by r_{pb}. Simple linear (non-regression[10] based) correlations between nominal level variables, and correlations between a nominal level variable and an ordinal level variable are covered under *Chapter 11 Chi-square*. Other associations between quantitative variables such as I-R and multicategorical nominal are infrequent and not covered in this chapter. They can however be conducted with regression calculations (Chs. 8 and 9).

These associations are all *bivariate correlations* also known as *zero-order* correlations. By far, the Pearson is the most common. Below is an explanation of the formula for the Pearson.

$$r = \frac{\Sigma (x - x)(y - y)}{nS_x S_y}$$

Equation 7.1 *Pearson Product-Moment Correlation*

Notice the similarity to the formula for the variance of a single variable. Excluding the standard deviations in the divisor the only other difference is whether you multiply the deviation score times itself for the numerator (variance for **x**) or multiply together the deviation scores for two IR variables (covariance for the Pearson **r**).

$$\text{var}(s^2) = \frac{\Sigma (x - \overline{x})(x - \overline{x})}{n} \qquad \text{cov} = \frac{\Sigma (x - \overline{x})(y - \overline{y})}{n}$$

Formulae 7.2, 7.3. *Comparison of Variance of X and Covariance of X and Y*

---

10) Multiple and bivariate regression approaches are covered in Chs. 8 and 9. No need to concern yourself with this until we get there.

The value of any lowercase **r** ranges from −1 to +1 with −1 meaning a perfect negative correlation and +1 meaning a perfect positive correlation. Look at the figures below and on the next page for the *Pearson r*. In each case two variables are plotted and the coordinates for the each subject are represented by a point. This is called a *scatterplot*. In the case of the positive and negative examples, the line of best fit is drawn through the plot indicating the direction or slope of the correlation, positive or negative. In the third example there is no correlation so **r = 0.** So for the positive association "As one variable increases so does the other", "The more counseling sessions a client attends the higher their self-image". In the case of the negative: "The more counseling sessions a client attends the less arguments he or she gets in". It doesn't matter which variable you make **x** and which you make **y**. They just coexist so the axes are interchangeable. We will see that in other statistical tests where one variable is hypothesized to depend on another variable that it does matter which variable is on which axis but not here.

One important feature of the Pearson is the assumption that each I-R variable is normally distributed, not skewed etc.

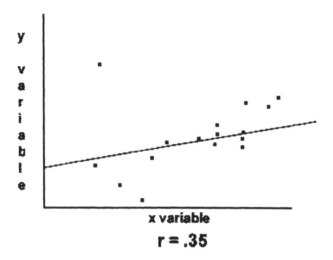

Figure 7a. *Scatterplot with Positive Association*

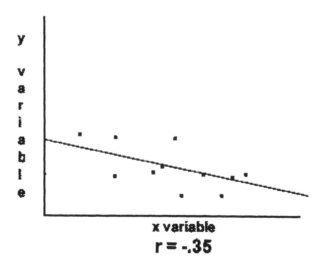

Figure 7b. *Scatterplot with Negative Association*

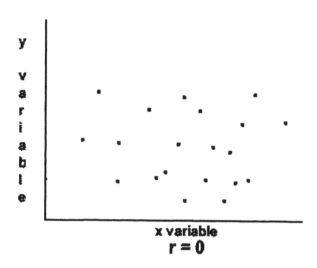

Figures 7c. *Scatterplot with Zero Correlation*

Let's use another example to demonstrate the Pearson **r**. Suppose you want to see if there is a statistically significant association between AGE and WEIGHT among a group of ten children (N=10). You have the data from Table 7.1, below.

NAME	AGE YEARS	HEIGHT INCHES			
	x	y	x^2	y^2	xy
Joe	12	65	144	4225	780
Jane	12	68	144	4624	816
Harpo	5	50	25	2500	250
Cindy	6	54	36	2916	324
Zeppo	5	48	25	2304	240
Chico	7	59	49	3481	413
Maria	8	62	64	3844	496
Sally	10	58	100	3364	580
Moe	7	55	49	3025	385
Curly	3	36	9	1296	108
Σ	75	555	645	31579	4392

Table 7.1 *Correlation Data*

Calculate the test statistic, in this case **r**, you then set up the null and alternate hypotheses[11]. In this case since your alternate hypothesis is that there is an association or correlation between age and height among children your null would be that there is not. First, the range of the test statistic for the three measures of association in this chapter **r**, r_{sp}, r_{pb} is between -1 and $+1$ (from $r = -1.00$ to $r = +1.00$). If you get something else, then you did your calculations incorrectly. No association at all would be **0**. Since this is an inferential test or a test of the null as it applies to the population, we use parametric symbols. **ρ** (rho) is the symbol for a correlation in the population just as **μ** is the symbol for the mean of a population. The null is **ρ = 0,** not **r = 0.**

- The null H_0: $\rho = 0$,   no correlation in the population
- The alternate H_1: $\rho \neq 0$,   some non-zero correlation in the population .

As mentioned it doesn't matter which variable you make **x** and which you make **y**.

The *null*, which you want to reject with at most a 5% chance of being wrong *is* that *there is no association.* The *alternate is* that *there may be an association.* Then you use the formula for the Pearson (Pearson Product-Moment Correlation) which is:

$$r = \frac{\Sigma \, (x - \bar{x})(y - \bar{y})}{n S_x S_y}$$

Equation 7.1 *Pearson Product-Moment Correlation*

Some use the *calculating version* with the components broken down thus:

$$r = \frac{\Sigma XY - \frac{(\Sigma X)(\Sigma Y)}{n}}{\sqrt{\left(\Sigma X^2 - \frac{(\Sigma X)^2}{n}\right)\left(\Sigma Y^2 - \frac{(\Sigma Y)^2}{n}\right)}} \quad {}_{12}$$

Equation 7.2 *Pearson Product-Moment Correlation (Calculating version)*

Substitute the values from Table 7.1.

---

11) If all you wanted to do was to measure the correlation in the sample you could stop here. But, we want to infer something to the population so we reject or do not reject the null.

12) This is the calculating formula as opposed to the descriptive formula given first. Either may be used.

$$r = \frac{4392 - \frac{(75)(555)}{10}}{\sqrt{\left((645) - \frac{(75)(75)}{10}\right)\left((31579) - \frac{(555)(555)}{10}\right)}}$$

Which will yield

$$r = .91.$$

So here is a strong positive association or correlation between AGE and HEIGHT in our sample. But we will always get some value different than zero. The real question is, *Is this correlation or simple bivariate association statistically significant?* In other words, *Can we reject the null of no correlation in the population?* There's a way to determine that, by using *Table B* in the back of the text. Part of *Table B* is reproduced next.

Significance Two-Tailed Test	p=.10	p=.05	p=.02	p=.01
df	r =	r =	r =	r =
1	.988	.997	.9995	.9999
2	.900	.950	.980	.990
3	.805	.878	.934	.959
4	.729	.811	.882	.917
5	.669	.754	.833	.874
6	.622	.707	.789	.834
7	.582	.666	.750	.798
8	.549	.632	.716	.765

Table 7.2, *Partial reproduction of Table B Pearson correlation*

Check on *Table B.* in the following way.

First calculate the *degrees of freedom*[13] for a Pearson correlation (df = N-2). So since in our problem N=10, our df =8

---

13) *Degrees of freedom* (df) is a bit difficult to understand. For now just know that it's an unbiased version of **N**.

1. Second move across from the position where $df = 8$ to the point where it intersects with your pre-determined alpha, usually .05.
2. Finally, if the **r** that you calculated is as least as large as the value of interest in the table then you may reject the null with a 5% chance of making a Type I Error. In our correlation between age and height, our value of .91 is statistically significant at at least .05 or less since .91 is equal to or larger than .632 in the table.

Spearman and Point Biserial Correlations

We have just learned the general characteristics of the Pearson **r** (zero-order correlation). And as you know this is a measure of association for two continuous variables (number line data). But what do you do when both variables are not interval-ratio? Well if they are both ordinal you may use either the chi-square test or the Spearman Rank Correlation. We will get to chi-square later. The formula for the Spearman is:

$$r_s = 1 - \frac{6 \sum d^2}{n(n^2 - 1)}$$

Equation 7.3 *Spearman Rank Correlation*

In this case the symbol *d* is the distance between a subject's score on each of two ordinal variables.

The point biserial (r_{pb}) correlation is simply the correlation between a continuous variable like 'height' and a dichotomous nominal variable like' sex'. It's not commonly found but it is a legitimate correlation for two variables. Often researchers will use a related measure of association **R** to achieve the same results as the point biserial. A later section will cover bivariate regression and the *big **R**.*

Shared Variance and Error

There's an additional aspect when two or more variables are correlated. It's the concept of *variance*, or *shared variance* if you will. The relationship of the Pearson r to the symbol for the *[correlation] variance* r^2 is the same as the standard deviation s is to the *variance* s^2 for a single variable. Here is an example of *shared* or *explained* variance. Just square the value of **r** to get the variance[14] for the correlation. So,

1) If

$r = +.70$, $r^2 = .49$ (i.e., 49% shared or explained variance)

2) If

$r = -.70$, $r^2 = .49$ (i.e., 49% shared or explained variance)

You may state this as *49% of the variance is explained by the data.* Since you are squaring a value of r, whether the correlation is positive or negative, the shared variance is always positive and ranges from 0 to 1.00, or 0% to 100%. *Shared variance* is the amount of variance accounted for by *both variables.*

---

14) *Variance* is the short term for *explained* or *shared* variance. Don't confuse it for the *variance* of a single I-R variable, a somewhat different use of the term.

Remember you take in to account the deviation scores for each variable. So instead of simple variance with one variable the variance for two is shared. But it may be called either *shared variance* or simply *variance.*

This also implies that some of the variance is not shared or is unexplained. By simply subtracting the variance from one, you can calculate the unexplained variance. For the example above,

$$(1 - r^2) = (1 - .49)$$
$$(1 - r^2) = .51$$

51% of the variance is not explained by the data. It is explained by something else. Either there are other factors or forces not measured by the two variables, or your measurements aren't reliable or valid enough, i.e., not precise enough. This second possibility is termed measurement error.

Unfortunately a simple bivariate correlation and shared variance doesn't really tell us much. The real world is more complex than a simple association between two variables. In addition we often want to consider the possibility of some sort of explanatory relationship between, three, four or more variables. The Pearson, Spearman, and Point-Bi-Serial correlations can't do this. We need something more to determine if chickens come before eggs. Or is it the other way around? At any rate, *linear regression* also known as *multiple regression*, or sometimes just *regression* is the way to go. It's presented in the next chapter while we cover bivariate regression in this chapter.

Causality

Perhaps the single biggest misunderstanding about statistics among the general public and beginning students is that of *causality* versus *correlation* or *association*. Let's differentiate the two concepts.

First, causality or a causal relationship states that one variable depends on another one. So a jail sentence may depend on the nature of the crime or weight loss may depend on the number of calories consumed or the amount of exercise. It so happens that causality or a dependent relationship between two variables cannot be confirmed with a statistical test. [15]

The particular tests we are examining in this chapter are simple correlations (either Pearson or Spearman below). A bivariate correlation describes how two variables change or vary with each other, i.e., co-relate. For example "The older you get, the slower you get", "The faster you run the higher your heart rate", "The more alcohol you consume, the slower your reaction time" etc.

So then in order to confirm causality or a causal relationship, three (or sometimes four) criteria are necessary:

1. There is a statistically significant test or association, as above
2. The presence of the time factor, i.e., that some event or situation precedes another (*x* comes before *y*) and,
3. The most difficult criteria, ruling out all other explanations or third variables.
4. The causal association must be replicated across different studies, different designs, and over time with various studies (this is a practical criteria that does not address causality per se as 1,2,3 above do)

---

15) By the way even under the best of conditions causality is almost impossible to prove in social science and not that easy in medicine either.

A simple Pearson correlation only meets the first criteria. Due to its formula it is impossible to determine 2, 3, or 4. This will become more clear once you understand the idea of a statistical association.

Let's look at an example that should easily explain the difference between causality and correlation. Suppose you are a demographer counting the number of births.[16] You find that there is a positive association between the number of storks and the number of babies born. The more storks, the more babies. Or, as one variable increases so does the second one. Since we know that babies are delivered by storks, this is no surprise. Is it? Actually only a child, a primitive, a network newscaster or some college professors would believe this.

So if babies aren't delivered by storks then how would you explain this factual correlation? Well the correlation between babies and storks is what's known as a spurious association. It is indeed a real correlation but a meaningless one since the true correlation is with a third variable, that of population. The greater the population, the greater number of buildings and the more storks you have. Also the greater the population the more births you have. The correlation between storks and babies is an illusion only. And no matter how strong or logical the correlation is, you may *not* impute causality. You may not state that babies are caused by stork delivery. In fact you may not even state that the number of babies or the number of storks depends on the population size. The idea of causality or dependence is inapplicable to a bivariate correlation. Period! When we believe that one variable actually depends on another we must use a different formula. We will get to this later.

For a social work example, suppose you want to see if there is some connection between the number of social work interns and the number of successful client placements in a job training program for the developmentally disabled across fifteen agencies. You do not have access to agency records so you are in the dark as regards the history of each agency and the time over which the placements occurred. Instead you have a hunch (more correctly the alternate hypothesis H_1:) that the more interns, the more placements. You can't test for causality without the time factor, as we stated. That is, you definitely may not say that a higher number of interns causes a greater number of successful placements only that there is an association or correlation. But you can meet one of the requirements for causality, that of statistical significance. You may test to see if there is a statistically significant association or correlation between the number of interns (we'll call this variable x) and the number of successful job placements (we'll call it y). There's a formula and symbol for this particular bivariate correlation as we shall see next.

Summary

1. One family of inferential tests is named *measures of association* or *correlations*. It means as one variable changes so does a second variable.
2. A correlation between two continuous variables, the test is called a *Pearson Correlation*.
3. The assumption for the Pearson correlation is that each variable is normally distributed.
4. A correlation between two ordinal variables the test is called a *Spearman Correlation*.
5. Correlations depend on the degrees of freedom and strength of the association.
6. Unexplained variance is due to third variables or measurement error
7. Causality cannot be detected by a correlation.

_____

16) Demography is the study of variables related to populations. *Demos* is from the Greek meaning *city*. The U.S. census is an example of demography.

Chapter Seven - Practice Problems

1. Can you correctly use a Person correlation to say that scores on a family functioning scale are associated with age of a parent?
2. Which measure of association would you use to detect a significant correlation between *sex* and *height*? What are the levels of measurement?
3. Can you correctly use a Pearson correlation to say that scores on a family functioning scale (range 20-80) depend on age of a parent (in years)?
4. Calculate the two-tailed Pearson correlation for the following measures of 'days truant from school' and age for a group of fourteen high school students. Is it significant at .05?

5. NAME	6. AGE	7. DAYS TRUANT
8. Seth	9. 14	10. 4
11. Sean	12. 16	13. 4
14. Rusty	15. 15	16. 8
17. Mariam	18. 14	19. 7
20. Hal	21. 16	22. 7
23. Phil	24. 18	25. 20
26. Suzie	27. 18	28. 12
29. John	30. 18	31. 10
32. Paul	33. 15	34. 8
35. Robert	36. 16	37. 9
38. Bill	39. 14	40. 0
41. Thomas	42. 17	43. 5
44. Chuck	45. 16	46. 0
47. Joanie	48. 15	49. 4

Chapter Eight
Bivariate Regression

Bivariate regression is a way to describe *a relationship between a single continuous variable and another variable* of any level of measurement. Since we have been talking about two I-R variables in the Pearson we can stay with that for now. Like the Pearson, bivariate regression may be used to correlate two I-R variables but also may be used to test whether the dependent variable (**y**) DEPENDS on the independent variable (**x**). You can't do this with the Pearson. The Pearson is only for correlations. In regression the **x** variable may be of any level of measurement. The assumption for the model is that of multivariate normality. This means that all the variables together make one single normally distributed curve. The equation for bivariate regression is:

$$\hat{y} = a + bx + \varepsilon$$

Equation 8.1, Bivariate Regression

1. A single dependent variable (DV)(must be continuous) [notice the '^' above the 'y'. This means 'estimated' and is pronounced 'hat y']	$\hat{y}$
2. A constant (the point where the line of best fit intersects y	**a**
3. An independent variables (IV) any level of measurement	**x**
*4. The coefficient for each independent variable (i.e., the slope on **y**)*	**b**
5. An error term for unexplained variance[17]	**ε**

The variance on **y** is estimated by the model (similar to explained or shared variance in the Pearson). The standard phrasing is "estimated variance on *y* ". Also the correct phrasing for stating the formula is that "*y* is regressed on ....".

---

17) Same as unexplained variance in a Pearson, Spearman, or Point Biserial *r*, but in this case it's $[1- R^2]$)

Once the coefficients for each IV and the constant are established, regression can be used to predict **y** by simply plugging in values for **a** and **x**.

Note that even though **y** is estimated from **x** and its coefficient **b**, the symbol we use in regression is R^2 (interpreted in the same way as r^2). In regression you're not particularly interested in **R** whereas in the Pearson you are interested in **r** more than r^2. But variance is variance in both cases. You can compare simple correlation with bivariate regression in Table 8.1 below.

	Pearson Correlation	**Bivariate Regression**
Symbol	**r & r^2**	**R & R^2**
Range of value	-1.00 to 1.00 for r, > 1.00 for r^2	>1.00 for both
Sig.	<=.05	<=.05
Y variable	I-R	I-R
x variable	I-R	any LOM
Uses	correlation	correlation or dependence

Table 8.1 Comparison of Pearson and Bivariate Regression

Let's use the regression model in an example. You might hypothesize that the DAYS TRUANT in your community is associated with or dependent on AGE OF TRUANT. In regression your **DV** or **y** must be interval-ratio data. You have to use some other test if it's not. The independent variable x can be any level of measurement in regression (although it is usually I-R or dichotomous nominal in social work research. For our example will assume the AGE OF STUDENT is coded in years (I-R variable). See Table 8.2 (next page) for the data on 15 subjects. As mentioned the IV may be of any LOM although it's usually I-R or dichotomous nominal.

The null and alternate hypotheses symbols are similar to the Pearson:

$$H_0: R = 0$$
$$H_1: R = 0$$

In our model NUMBER OF DAYS TRUANT may be tested as either a) correlated with or b) dependent on AGE OF STUDENT. This is the beauty of regression and its advantage over a simple Pearson. Another advantage as mentioned is that the IV may be of any level of measurement. (See *Table 8.4*, p.56 for raw data)

Substituting our variables in the regression equation we get:

$$\hat{y} = a + bx + \varepsilon$$

where **y** = NUMBER OF DAYS TRUANT
where **a** = a constant for the equation
where **x** = AGE OF STUDENT
where **ε** = the error term (unexplained variamce and noise)

Data in Table 8.3 are used to illustrate a bivariate regressionform an SPSS printout. Notice that in the *Model Summary* that the correlation is .597. If you ran a Pearson correlation you would get the identical value for **r**. The **R²** is .396 but the Adjusted **R²** of .306 is the important one since it takes into consideration sample size. You may state from this that *30.6% of the variance is explained by the model, or 30.6% of the variance is explained when regressing NUMBER OF DAYS TRUANT on AGE.* Thus the unexplained variance is 69.4% (100% - 30.6%, i.e., the error term[18]). No one actually calculates regression by hand but it's good to understand some of the process.

Model Summary

Model	R	R Square	Adjusted R Square	Sig.
1	.597	.356	.306	.003

a Predictors: (Constant), AGE
b Dependent Variable: DAYSTRUA

Table 8.2. Regression Analysis Results

Here's another example. Let's say you want to explain or predict variance on the SCORES ON THE PSS (Parent Satisfaction Scale) with the NUMBER OF CHILDREN in each family used in the investigation. The formula is the same: **y = a + bx + ε**

where $\hat{y}$ = SCORES ON THE PSS (DV)
where **a** = a constant for the equation
where **x** = NUMBER OF CHILDREN (IV)
where **ε** = the error term (unexplained variance)

However the magnitude of **Y, a** etc. as well as the sig. will of course be different. In the next chapter we will cover multiple regression which is just an extension of bivariate regression. If you have followed this chapter then multiple regression should not be too difficult.

---

18) The *error term* does not mean you made a mistake. It is comprised of variance that we don't have variables for or measurement error due to imprecise scales.

NAME	AGE	DAYS TRUANT
Mitchell	14	7
Seth	14	4
Sean	16	4
Rusty	15	8
Mariam	14	7
Hal	16	7
Phil	18	20
Suzie	18	12
John	18	10
Paul	15	8
Robert	16	9
Bill	14	0
Thomas	17	5
Chuck	16	0
Joanie	15	4

Table 8.3 Truancy Data

Summary

1. Regression can be used to correlate variables as the Pearson does, or to hypothesize that **y** depends on **x**, i.e. a causal or dependent relationship.
2. The DV must be I-R but the IV's may be of any LOM.
3. The assumption is that together both the DV and IV form a normal distribution, i.e., multivariate normality.
4. P-values are interpreted in the usual way.

Chapter Eight – Practice Problems

1. If R = .30, what is the unexplained variance?
2. If r = .30, what is the unexplained variance?
3. Name two differences between a simple Pearson correlation and a bivariate regression?
4. Suppose you want to correlated age and sex, which measure of association would you use?
5. What is 'a' in the regression equation?

Chapter Nine
Multiple Regression

Basics

Multiple regression follows the same logic as bivariate regression but has more than one independent or **x** variable[19]. Both are known as *linear regression* since they both use the formula for a straight line. In the real world there is rarely a single bivariate model that explains a phenomenon satisfactorily. There are always other factors or forces in play. In multiple regression you can add third, fourth, fifth, etc. variables and detect the change in shared or explained variance. For example FAMILY FUNCTIONING needs more than a single variable such as MARITAL STATUS or NUMBER OF CHILDREN to adequately explain its variance. So you can look beyond the simple 2-variable correlation to see what else could account for unexplained variance $(1 - R^2)$.

Like bivarite regression the assumption for the model is that of multivariate normality. This means that all the variables together make one single normally distributed curve. The equation for linear regressions is an extension of bivariate regression where:

$$\hat{y} = a + b_1x_1 + b_2x_2 + b_3x_3 + \ldots + b_ix_i + \varepsilon$$

Equation 9.1 Multiple Regression

1. *A single dependent variable (DV)(must be continuous) [notice the '^' above the 'y'. This means 'estimated' and is pronounced 'hat y']*	$\hat{Y}$
2. *A constant (the point where the line of best fit intersects y*	**a**
3. *Independent variabless (IVs) any level of measurement*	$x_{n\,(1\ldots i)}$
4. *The coefficient for each independent variable (i.e., the slope on $\hat{y}$)*	$b_{n\,(1\ldots i)}$
5. *An error term for unexplained Varianc*	ε

---

19) Some incorrectly use the term *multivariate*. Multivariate tests have more than one DV not more than one IV as does multiple regression and other tests such as multi-way and factorial ANOVA. Regression and the like are *univariate*.

Let's use the regression model with an example from *Chapter Eight* but we'll add another **x** variable in accounting for variance on **y** and in prediction of ŷ. So again as before:

FREQUENCY OF CHILD ABUSE = (1) PARENTS' HISTORY OF ABUSE + (2) FAMILY INCOME

Substituting our variables in the regression equation we get:

$$\hat{y} = a + b_1x_1 + b_2x_2 + \varepsilon$$

where $\hat{y}$ = FREQUENCY OF CHILD ABUSE

where **a** = a constant for the equation

where x_1 = PARENTS' HISTORY OF ABUSE

where x_2 = FAMILY INCOME (I-R)

where ε = the error term (unexplained variance)

Here's another example, a 'measurement' study using published scales measuring various types of fatigue. Let's say you want to explain or predict variance on the GENERAL FATIGUE SCALE with scores on the DEPRESSION SCALE , VITALITY SCALE, and AGE. The formula is the regression formula with three independent variables.

$$\hat{y} = a + b_1x_1 + b_2x_2 + b_3x_3 + \varepsilon$$

where $\hat{y}$ = GENERAL FATIGUE

where **a** = a constant for the equation (**y** intercept)

where x_1 = DEPRESSION SCALE

where x_2 = VITALITY SCALE

where x_3 = AGE

where ε = the error term (unexplained variance)

First let's see how this regression analysis is different from simply correlating the General Fatigue Scale with the others. Compare separate Pearson correlations between General Fatigue and the other scales with the correlation matrix (Table 9.4).

	GEN FA	DEP	VIT	AGE
GENFAT Pearson Correlation Sig. (2-tailed)	**1.000**			
DEPRES Pearson Correlation Sig. (2-tailed)	**.610** **.001**	1.000		
VITALIT Pearson Correlation Sig. (2-tailed)	**.568** **.001**	.747 .000	1.000	
AGE Pearson Correlation Sig. (2-tailed)	**.526** **.001**	.608 .000	.721 .000	1.000

Table 9.4 *Pearson Correlations*

GENERAL FATIGUE and DEPRESSION SCALE

$$r = .610, \quad r^2 = .372$$

GENERAL FATIGUE and VITALITY SCALE

$$r = .568, \quad r^2 = .323$$

GENERAL FATIGUE and AGE

$$r = .526, \quad r^2 = .277$$

Total $r = $ ~~1.724~~ $r^2 = $ ~~.972~~?

The total **r** for the three associations is .1.724 which is impossible The total **r²** for the three associations is .972 which is highly unlikely. You can't just add up a bunch of bivariate correlations to see how they all work together. You have to use linear regression if you want to account for them all at once. And this is the thing with multiple regression: ALL AT ONCE. Now let's break this down with multiple regression. There are three independent variables. We want to do two things.

1. See how much total variance there is when we enter them all at once.
2. See how each one contributes a piece of the total variance. We do this in steps, each time adding another ingredient if you will.

Step 1. GENERAL FATIGUE and DEPRESSION SCALE

The variance (R square) is .357 or 36% (See Table 9.6)

Step 2. GENERAL FATIGUE and (DEPRESSION SCALE and VITALITY SCALE)

On Step Two you entered the second independent variable, the DEPRESSION SCALE and the VITALITY SCALE . So they're both in there. Your combined explained variance is (**Adj R^2** = .372)[20], 37.2%, an increase of 15% above the DEPRESSION SCALE alone. See Table 9.5 for details of the steps.

Step 3. GENERAL FATIGUE and (DEPRESSION SCALE and VITALITY SCALE and AGE)

On Step Three you keep the first two independent variables and add a third, AGE. Adding this increases total explained variance by 7% more. The **final Adj. R^2 is .374** so 37% of the variance is accounted for by all the steps added up. When you subtract this from 100% then the error is 63%, i.e., 63% of the variance on **y** is not explained by your three independent variables. Sometimes *error* is termed *noise*. A modified SPSS regression printout of these data is shown in Table 9.5.

	Model Summary			
Model	R	R Square	Adjusted R Square	R Square change
1	.610	.372	.357	0.00
2	.633	.401	.372	0.15
3	.646	.417	.374	0.02

a Predictors: (Constant), **DEPRESSION**
b Predictors: (Constant), **DEPRESSION, VITALITY**
c Predictors: (Constant), **DEPRESSION, VITALITY, AGE**
Remember **a** is the constant and comprises part of the explained variance.

Table 9.5 *SPSS Regression Results*

---

[20] Adjusted R^2 is the usual R^2 adjusted for sample size. It is interpreted in the same way and more accurate.

In terms of interpretation you can see that adding independent variables increases the explained variance on your dependent variable above entering them separately, one at a time. This is more in the reality of phenomena. Most phenomena in nature are multifactorial.

The above explanation of multiple regression is minimal. Much more can be included but for simplicity's sake wasn't. For example like bivariate regression the IV's may be of any LOM or combination of LOM's. In addition adding second or more variables may do something besides increase explained variance. It may do nothing at all. The variables may interact with each other and also have *interactional effects* on the DV in the form of *partial* or *semi-partial* (part) relationships as they are known.

There are a number of diagnostic applications associated with regression. One is *multicollinearity*. This means that two or more independent variables may be highly correlated with each other which can hide their correlation with the DV. Other tests such as *tolerance* can detect the limits of the model.

Summary

1. Multiple regression is much like bivariate regression in that the same liner formula is used.
2. The assumption of multivariate normality applies.
3. Both have a constant; both regress **y** on at least one **x**; both yield explained and unexplained variance.
4. Multiple regression has more than one independent variable.
5. Multiple regression has the advantage of more precise assessment of variance by examining the variance change after adding independent variables.

Chapter Nine: Multiple Regression – Practice Problems

1. Name two uses for multiple regression.
3. How might you increase explained variance?
4. What is the symbol for unexplained variance and what is it?
5. What are the LOM's in a regression analysis?

Chapter Ten
t-Test and One-Way Analysis of Variance

Before we get to the details of the inferential tests presented in this chapter you need to understand the general differences between *measures of association* and *tests for group independence*. So far the inferential tests we have covered are all measures of association or correlations of one type or other. These are **r, r$_s$, R** that describe linear relationships between two or more variables. As **x** changes so does **y**. As AGE increases so does HEIGHT.

Now we're talking about tests for group independence, the other broad family of inferential tests. They are quite different from correlations. They are NOT correlation measures or measures of association like **r.** So what are tests for group independence? If you examine *Table 10.1* you can see the general differences between measures of association and tests for group independence. The short version is that **T** and **F** test for significant differences between the means of two or more groups but do not detect correlations. **r** etc. can only detect correlations.

Measures of Association	Tests for Group independence
1) **r, r$_s$, R**	1) **t, F**
2) Correlations, statistical association 3) no dependency 4) causality, no 5) **R** is a strange exception where you may if you wish say that **Y** depends on **X**, at least statistically. Even so, **R** is still a measure of association regardless of its use)	2) Actual sig. differences in means or cell count depending on which group you are in 2) no correlation 3) causality, yes
6) You can only say **X** is associated with **Y** or is correlated with **Y**	6) You state that **X** (IV) depends on **Y** (DV)

Table 10.1, *Differences Between Measures of Association and Tests for Group Independence*

This second family of tests for group independence essentially tests to see whether there is a significant difference in the means or in the case of nominal data, the modes of two or three or more groups Comparing the class average of three sections of introduction to sociology is an example. Testing to see whether BLOOD PRESSURE depends on whether you are in the exercise GROUP or the non-exercise GROUP is another (example below). Both of these examples require that the dependent variable is interval-ratio.

The *dependent variable* BLOOD PRESSURE depends on which group the subjects are in and GROUP membership is the *independent variable* (IV) upon which the (DV) depends, such as in the diagram, next:

DV ← ← ← **IV**

BLOOD PRESSURE   GROUP (coded **1,2** or perhaps **1, 2, 3**)

Must be interval ration   Must be nominal for t-test and one-way

Depends on …..

Now for the specifics of the t-test and one-way ANOVA.

t-Test

In the last two chapters we learned most of the important measures of association found in social work research. We covered the association or correlation between two continuous variables, two ordinal variables and a single continuous and a single dichotomous nominal variable. That is, we learned bivariate (zero-order) correlations when no dependency is presumed. Finally we went on to a type of association when dependency is assumed, that is, some **y** depends on some **x** or **x's** (in other words single or multiple independent variables). When the dependent variable *y* is continuous (and it often is) the analysis of choice for correlations is typically linear regression.[20]

The other family of inferential tests is *tests for group independence*. In Chapter Seven we were actually looking at a simple version of a test for group independence when we learned the null and alternate hypotheses. Much more often than not we're not too concerned which groups is significantly larger than the other, only that the two are significantly different. We can always inspect the two means and compare them. In addition we usually want to detect significant differences possible in either direction. We could miss something important if we put the whole 5% in one tail and then tested a directional hypothesis. Better to use a two-tailed or non-directional test where alpha is split between the two tails with 2.5% in each tail. We only use a directional one-tailed test when we are absolutely sure that if there is significance it will be in the intended direction, or if opposite findings are unimportant. More on this later.

So for the two-tailed, non-directional tests the hypotheses are as follows:

$$\mathbf{H_0}{:}\mu_1 = \mu_2$$
$$\mathbf{H_1}{:}\mu_1 \neq \mu_2$$

This null is actually the null for our first new equation, the *t-test*. The term *test for group independence* is derived from the null. That is, when the null is true then there is no difference between the means or, the continuous variable is *independent* of group membership. This test of two groups is called the *t-test*.

---

20) We can also simply test whether one group is significantly different from another group without using a correlation or linear formula (**R** actually tests the slope for 'group membership').

When you want to directly test (as opposed to testing the slope in regression) whether some continuous variable *depends on* a *dichotomous nominal level independent variable*, you typically use the *t-test*. Again the DV must be continuous and the single IV must be dichotomous nominal. No other variables are allowed, so it has its limits. Except for one new symbol, the *standard error of the difference* (in means), you know the elements of the formula already. The numerator is obvious.

$$ t = \frac{\bar{x}_1 - \bar{x}_2}{S_{\bar{x}_1 - \bar{x}_2}} $$

Equation 9.01, *t-test*

Just subtract one mean from the other. It doesn't matter which one. But the divisor is new. It's called the standard error of the difference in means or just the *standard error of the difference*. It has its own formula which you must calculate first before using it as the divisor

$$ S_{\bar{x}_1 - \bar{x}_2} = \sqrt{\frac{s_1^{\,2}}{n_1} + \frac{s_2^{\,2}}{n_2}} $$

Equation 9.02 *Standard Error of the Difference in Means*

The *standard error of the difference in means* (S.E.D.) is simply the square root of the sum of each standard deviation divided by the **n** for each category of '**x**' (i.e., **x₁**, **x₂**). The null and alternate hypotheses you already know. Remember that the null and alternate apply to population inference and hence are in the form of Greek letters. For the t-test, to repeat, the correct null and alternate hypotheses are:

$$ H_0 : \mu_1 = \mu_2 $$
$$ H_1 : \mu_1 \neq \mu_2 $$

In plain English the null for the t-test says "there is no statistically significant difference between the means of two groups in the population from which I drew my sample". Rejecting the null entails some chance of making a Type I error of course. We'll get to obtaining p-values shortly.

Let's see an example. Suppose you wanted to compare the success rates for two family reunification agencies. At Agency A, the staff uses conventional social work practice techniques. We'll call this the 'Control Group (1)'. At Agency B, the staff has a longer but more intensive preparation. We'll call this the 'Experimental Group (2)'. These are the two attributes of the independent variable (IV), group membership (coded 1, 2). Figure 9.01 shows both groups on the same curve with means and standard deviations that will be hypothesized in the population from which you drew your sample.

You hypothesize that the percentage of families successfully reunified *depends on* which of the agencies that the families are involved with. Family reunification is the dependent variable (DV). Again in this case family reunification is hypothesized (the alternate hypothesis) to depend on group membership coded (1, 2).[21]

---

21) The values of the two groups are arbitrary. You could just as correctly code them as 0 and 1.

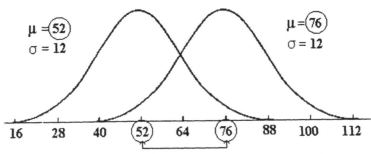

Figure 9.01 *Means of two groups (Left side, control group; Right side experimental group)*

<table>
<tr><td colspan="2" align="center">**Two agencies non-directional t-test.**</td></tr>
<tr><td>*Agency A*
(control group = 1)</td><td>*Agency B*
(experimental group = 2)</td></tr>
<tr><td>N = 11</td><td>N = 10</td></tr>
<tr><td>X_1 = 52% success rate
(SD 12.0)</td><td>X_2 = 76% success rate
(SD 12.0)</td></tr>
</table>

1. First you go ahead and calculate the standard error of the difference in means using the formula

$$S_{x_1 - x_2} = \sqrt{\frac{S_1}{n_1} + \frac{S_2}{n_2}}$$

$$S_{x_1 - x_2} = 1.51$$

2. Then just simply divide the difference in means by the standard error of the difference for the t-test formula.

$$t = \frac{\overline{x}_1 - \overline{x}_2}{S_{\overline{x}_1 - \overline{x}_2}}$$

$$t = \frac{52 - 76}{1.51}$$

$$t = -15.89$$

3. The final step is to identify the critical value for t. These are the cut points for the 95% C.I. Like the '**r**', it has degrees of freedom associated with it. **N** is the number of subjects in each group. For the t-test:

$$df = N_1 + N_2 - 2$$

So for our problem we have 19 degrees of freedom.

$$\mathbf{df = N_1 + N_2 - 2}$$

$$\mathbf{df = 10 + 11 - 2}$$

$$\mathbf{df = 19}$$

4. Then we use *Table C*, the Critical Values for **t** and see if our calculated absolute value of **t** = -15.89 is *at least as high* as the minimal value we need to reject the null at .05.[22] From the table, the minimal value needed to reject the null at .05 with 19 degrees of freedom for a non-directional, 2-tailed test (i.e., 2.5% in each tail) is 2.09. Since our value is -15.89 we can reject the null of "no statistically significant difference between the two groups among the population from which we drew our sample."

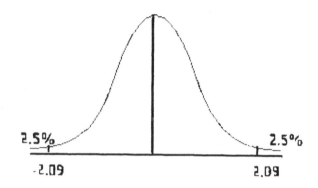

Figure 9.02 *Critical Values of 2-tailed T-test (df = 19)*

Directional t-test

As with the simple zero-order bivariate correlation **r**, the **t** can be directional or one-tailed too. That is, our 5% can be all in one tail as opposed to split between the two tails. It's still the 95% C.I., but just in one direction. When this is the case, just as with the *Pearson r*, it's easier to reject the null as long as your hypothesis is in the hypothetical direction. The null and alternate hypotheses for the *directional t-test* are:

$$\mathbf{H_0}:\mu_1 \geq \mu_2$$
$$\mathbf{H_1}:\mu_1 < \mu_2$$

So if you suspect that the mean of successful family reunifications for group 2, your experimental group, will be greater than the mean for group 1, your control group then the alternate will be in the suspected direction (>) and the null will cover all other possibilities (< or =). The other possibilities

---

22) The t-values are given in absolute values. For a non-directional test you are not concerned with which mean is greater than the other. Only that they are not equal. Therefore whether the t is positive or negative makes no difference.

are that the two means could be less than or equal to each other, as long as the mean of group two is not larger than the mean of group one (i.e., H_1:).

Let's take our same example of family interventions and make it directional (a one-tailed test that is). Well use the same data and hypothesize that our longer more intensive preparation group (experimental group, 2) will have a greater success rate in family reunification than our regular services group (control group, 1). This is clearly directional. We cannot reject the null if the mean for group one isn't first, greater than the mean for group two (the directional aspect), and second, in the correct direction.

Our degrees of freedom are still 19. So we go to the Critical Values Table again and look up the rejection point for a t of -15.89 with 19 degrees of freedom for the *one-tailed test*, not the first version, the two tailed column. The critical value of **t** is 1.73. Our calculated value is −15.89 is larger in magnitude than the critical value of 1.73.

Can we reject the null at .05? NO! Why not, you ask? Well if this were a *non-directional t-test* only the absolute value of **t** needs to be considered. In that case, that is our first case above, we can reject the But here not only do we consider the magnitude of **t** but also the direction since out hypotheses are directional. That is our null is rejected only if the mean of group one is larger than or equal to the mean of group two. But it isn't. In this case of the one-tailed test it doesn't matter how large **t** is if it's not in the right direction. Fortunately it is not usually necessary to use a *directional t*. The only infrequent times that we would use a *directional t* is when we have no doubt about direction *or* if findings in the opposite direction are unimportant. In social work practice we want to know if our interventions improve or possibly harm clients. Social workers can harm clients. It's not intentional but it does happen just as doctors can unintentionally harm patients. Since we ethically don't want to harm clients you're better off using a *non-directional t-test*. In this way we can look to both tails for the rejection region.

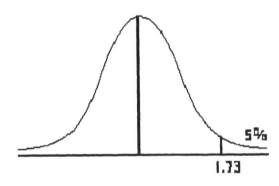

Figure 9.03, *Critical Value for one-tailed t (df = 19)*

One-Way Analysis of Variance

Now what do you do when you want to compare the means for three or four groups? The *t-test* is limited to a continuous DV and a dichotomous nominal variable only. That's it! No other variations are allowed. If you think about it, there's no room in the formula for anything else. Well if you want to compare the means for three or more groups and your DV is continuous, you use what is termed the *one-way ANALYSIS of variance (ANOVA)*.

We'll look at an example, but first let's see what the one-way actually does. It's a little different from what the t-test does. The t-test tests whether the means of two groups are significantly different. It does this by determining how many standard deviation units for the difference in means or S.E.D. the means are from each other, depending on sample size. It's pretty simple. The one-way ANOVA on the other hand technically tests to see whether variances for the attributes (groups) differ significantly from the whole. It does this by finding if its test statistic called **F** is significant or not.

Comparisons between the t-test and Oneway ANOVA

t-test	One-way ANOVA
Continuous dependent variable	Continuous dependent variable
Dichotomous nominal independent variable	Trichotomous or greater independent variable (can be dichotomous)
Assumes normal distribution of DV	Assumes normal distribution of DV
Can be directional or non-directional (tailedness)	Can only be non-directional
Test statistic is **t**	Test statistic is F ($t^2 = F$)
df, yes	**df,** yes

Table 9.2, *Comparisons Between t-Test and Oneway ANOVA*

variability between the groups in the hypothetical population. The null is that there is nothing significant going on. Table 9.01 on page 79 might help to understand this. In this example we are comparing the IV of family agency (Lutheran Social Services, Family Service League, Jewish Family Services) to detect whether there is a significant difference between groups in the number of foster care placements per month. Note that the DV is placements (continuous) and the IV is GROUP MEMBERSHIP (LSS, FSS, JFS).

The bottom line is that the *F-test* examines the patterns of the within group variances when compared to the between group variances. If this ratio is large enough then you can reject the null. Simply put, a significant **F** suggests that there is a significant difference between at least two groups in the population. We won't know which two groups or even if all three are significantly different from each other – only that at least two are significantly different. Each type of agency has its own variability within and that there is variability between groups as well.

The 'oneway' in *oneway ANOVA* means a single independent variable, group membership. In intermediate statistics (beyond the scope of this book) you will see that the multiway and factorial ANOVA are analyses that have more than one IV). The commonality in any type of test is the *analysis of variance* regardless of the number of independent variables. So the question is - How does the oneway ANOVA actually analyze variance?

First remember that the numerator of the variance formula is the sum of squares, i.e., the deviation score squared, $S(x-x)^2$. The ANOVA actually analyzes this variance or more precisely the relationship between the **SS** divided by the **df** (i.e., **N-1**). Now look at *Diagram 10.2*. In the diagram you can see on the vertical axis a representation of ALL the subjects. Besides being a member of the whole large group, each subject is a member of a subgroup or attribute (1 or 2 or 3).

So as I said, the ANOVA compares the pattern <u>within</u> (this is the variance **SS/df** for ALL THE MEMBERS all at once compared to the pattern <u>between</u> groups, that is the three variances **SS/df**, treating each subgroup as if it were a single person. Specifically it takes the ratio of the mean of squares between, i.e., the average for the sum of squares between groups and the mean of the squares within.

You can see that for three groups the variance within each group is indicated as well as the variance among all the groups. The **F** is the ratio of between groups vs. within (each) groups. You're testing **F** as to its significance.

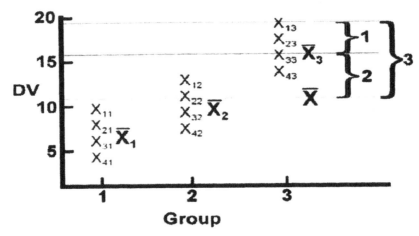

Diagram 10.1, *Variance Between and Within Groups*

$$F = MS_B / MS_W$$

Equation 10.4, *The F-test*

First we need to calculate the mean of squares (MS_B) between and the mean of squares within (MS_W). These are calculated by dividing the **SS** by **df** (Not actually **n** but rather degrees of freedom, an approximation of **n**). And in this case there are two places where we need degrees of freedom, between *and* within. The two sub-formulae then the

Mean of squares between	$MS_B = SS_B / df_B$
Where,	$df_B = k - 1$
And	k = number of groups.
Then, mean of squares within	$MS_W = SS_W / df_W,$
Where	$df_W = N - K$
And	N = total number of subjects for all groups.

But how do we get the SS_B and SS_W you ask? Good question. Remember even for the simple variance the **SS** is the quantity of a score minus a mean squared and added up. So we calculate first the *sum of squares total* (SS_T). To do this you add up the squares of each score no matter the group subtracted from the *grand mean*. The grand mean is the mean of <u>all</u> the scores no matter the group. That is you ignore the group and treat the total N as one big group.

$$SS_T = \Sigma(x-\bar{x})^2$$

Then you calculate the *sum of squares within* (SS_W) by doing a SS for <u>each group</u> as if you were doing three separate variance calculations. So now you have the SS_W to eventually divide by the df_w. Since you have the SS_T and now the SS_W you simply subtract the SS_W from the SS_T to get the sum of squares between or SS_B. Then you just finish out the calculation once you ascertain the degrees of freedom.

$$SS_W = \Sigma\,(x_1-\bar{x}_1)^2 + \Sigma\,(x_2 - \bar{x}_2)^2 + \Sigma\,(x_3-\bar{x}_3)^2$$

$$SS_B = SS_T - SS_W$$

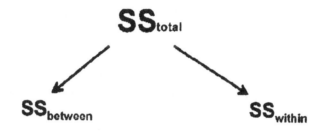

Diagram 10.2, *Sums of Squares Relationship*

Then the test statistic as we learned is:

$$F = MS_B / MS_W$$

If **F** is significant then you reject the null. But how do you find out if it's significant? Well, you just go to yet another table, in this case *Table D* in the back of your text as well as reproduced on the next page. This particular table only applies to an alpha of .05. The null hypotheses for a oneway ANOVA with only three groups are as follows:

$$H_0: \mu_1 = \mu_2 = \mu_3$$

Notice that there are two types of degrees of freedom. The top row or horizontal axis identifies the correct column for the df_B **(K-1)**. If you have three groups for example, **K-1 = 2df**. The far left column identifies the df_W **(N-k)**. This will depend on how many total subjects you have

So you identify the cell where these two degrees of freedom intersect and the value of **F** is the minimum value you need to reject the null at **.05.**

Most texts on statistics will have students calculate a simple oneway ANOVA by hand but I see no need for this. Rather let's just continue with our example above for the number of foster care placements dependent on type of family agency. We have three agencies, Lutheran Social Service, Family Service League, and Jewish Family Services. Our null hypothesis (as a test for group independence) is that the number of placements is independent of type of family agency. We calculate an **F** of **6.332.**

Part of Table D
Significance of F = 0.05
(confidence = 95%) for small samples
Vertical Axis = Denominator Degrees of Freedom (dDf)
Horizontal Axis = Numerator Degrees of Freedom (nDf)

df	1	2	3	4	5	9	14	19	24
1	161	200	216	224	230	240	245	248	249
2	18.5	19.0	19.2	19.2	19.3	19.4	19.4	19.4	19.4
3	10.1	9.55	9.28	9.12	9.01	8.81	8.71	8.67	8.64
4	7.71	6.94	6.59	6.39	6.26	6.00	5.87	5.81	5.77
5	6.61	5.79	5.41	5.19	5.05	4.77	4.64	4.57	4.53
9	5.12	4.26	3.86	3.63	3.48	3.18	3.02	2.95	2.90
14	4.60	3.74	3.34	3.11	2.96	2.64	2.48	2.40	2.35
19	4.38	**3.52**	3.13	2.90	2.74	2.42	2.26	2.17	2.11
24	4.26	3.40	3.01	2.78	2.62	2.30	2.13	2.04	1.98

Table 10.1 Partial table D (at .050)

But we still need to see if our F-statistic is significant. We go to the cell in Table D that represents our **df$_B$** (i.e., 2) and **df$_W$** (i.e., 19). This value is 3.62. Our **F** is rounded to 6.33. since this exceeds 3.62 we may reject the null at .05 for that particular cell. The question then becomes which two out of the three groups are significantly different. It so happens that there is a statistic that you may select when you run the oneway ANOVA that is named Tukey, after the statistian John Tukey. It's a post hoc test (not actually part of the ANOVA) that will tell you which groups are significantly different from each other. So far our dependent variable has been interval-ratio, but there are a number of times that you may have a correlation or association between two nominal or even ordinal variables. When this is the case, you use a statistical test called chi-square. That's the next chapter.

Summary

1. The test for significant difference between two means is the t-test.
2. The t-test may be directional or non-directional.
3. The t-test had degrees of freedom.

4. The test for significant differences among three or more groups is the one-way analysis of variance or ANOVA.
5. The ANOVA is non-directional.
6. The F-test had two different degrees of freedom to look up.
7. The Tukey test can detect significant differences among the means of two groups with a three or more groups F-test.

Chapter Ten – Practice Problems

1.   a) Calculate the value of t for the following data.
        Group A: 12, 3.3, 55, 21, 6, 7.7, 43, 51, 2
        Group B: 12, 3.3, 102, 23, 5.5, 57, 46, 2, 61
     b) Is the t-value significant at .05? What about .01?
2. Can you detect which group(s) is significantly different from another group(s) in the one-way?
        Why or why not?
3. Identify the correct test to use for the following study
        You believe that students in your program who take a second research course as an advanced elective have a higher GPA than do students who take an advanced elective in child welfare.

Chapter Eleven
Chi Square

The Pearson Chi-Square (x^2), while it may be used as a *measure of association* for categorical data is most often used as a *test for group independence* like **t** and **F**. That is, some outcome depends on which category or group you are in. The difference between this (*non-parametric[23]*) and the *parametric* tests that we have covered is that in addition to group membership the DV or outcome is also categorical (ordinal but most often nominal). The chi-square data is presented in crosstabs format. Three examples are:

1. Whether you get an A or a B depends on whether you are male or female in a 2 x 2 table
2. Does having high blood pressure (yes or no, 0 or 1) depend on race (1, 2, 3, 4) in a 2 x 4 table?
3. Does graduating H.S. (yes or no) depend on whether your parents graduated H.S. (yes or no) (2 x 2 table)?

The formula for chi-square is:

$$X^2 = \Sigma \frac{(O - E)^2}{E}$$

Formula 11.1 *Chi Square*

The **X** is not a capital **x** but rather a chi (pr. k'eye). **O** mean observed and **E** means expected. The O's and E's apply to each cell in the crosstab. The sigma simply means to add up for each cell.

Now look below at the X^2 printout from SPSS for the variable EDUCATION (with three attributes, 1, 2, 3) dependent on SEX (two attributes, 1, 2). The data reported in the cells of the 3 x 2 table below are the observed, **O**.

SEX * EDUCATION Crosstabulation[24]
Count

SEX	EDUC. grade	H.S.	college	Total
male	3	0	12	15
female	6	9	3	18
Total	9	9	15	33

---

23) Parametric or parallel to the meter is a test where the DV is I-R. A non-parametric test uses variables (esp. the DV) that are not parallel to the meter, i.e., categorical (ordinal or nominal). This is not to be confused with our earlier term parameters which means measures that apply to populations rather than samples.

24) I always use the columns for the dependent variable and the rows for the independent (column membership depends on row membership). The columns or rows may be reversed since the X^2 value will not differ either way. I keep the columns as the DV since it can be confusing when variables that are not obviously dependent or independent.

Chi-Square Tests

	Value	df	Asymp. Sig. (2-sided)
Pearson Chi-Square	**15.253**	**2**	**.000**
Likelihood Ratio	19.005	2	.000
Linear-by-Linear Association	6.717	1	.010
N of Valid Cases	33		

a 4 cells (66.7%) have expected count less than 5. The minimum expected count is 4.09.

Table 11.2. *Chi-square Print-out for a 3 x 2 Table.*

Let's see what these table results mean. In the first table the numbers in the far right column, 15, 18 and 13 are the row totals from left to right. For example 3 + 0 +12 = 15 males etc. Similarly adding down such as 3 + 6 = 9 for grade school graduates represent a column total. Both row and column totals are termed *marginals*. The grand total or N = 33 is in the bottom right cell.

In the second table we have 15.253 which is the actual calculated value of chi-square. .000 is the *Asymptotic Sig. (2-tailed)* or simply put the p-value. *df* means degrees-of-freedom. We'll get back to that. Don't worry about other cells.

OK. Now we'll look at the calculations. For each cell complete **(O-E)2 /E**. You are given the observed so all you need do is find out the E and simply do the arithmetic. But how do you know what E is?

Calculating Expected Values in Chi Square

The observed values are taken from the data you have collected. That's easy. Actually what you are testing is whether what you have (observed) is different enough from what you would expect if there was something going on. Of course it's what you would expect with some chance (5%) of being wrong when you reject the statement of 'no difference' between the expected and observed (null). So how do you get **E**? The expected values are calculated using the observed as a guideline. The expected values are a proportion of the observed adjusted for the total **N** and for N in the particular row and column in question. So looking at our table above ignoring the observed and keeping only the marginals we get:

SEX * EDUCATION Crosstabulation

Count

SEX	EDUC. grade school	H.S.	college	Total
male				15
female				18
Total	9	9	15	33

Table 10.3, *Crosstab with marginals only*

Now we calculate the expected for each cell by a simple method which is:

observed row total X observed column total = expected for a cell
total N

Don't get confused. We'll go through it exactly as it should be done We are calculating expected values based on observed values. First understand that each cell is an intersection of a row and a column so no two need be the same except by accident. We have six cells with observed values. We'll do them one at a time

	EDUCATI			Total
SEX	grade school	H.S.	college	
male	4.1 a	4.1 b	6.8 c	15
female	4.9 d	4.9 e	8.1 f	18
Totall	9	9	15	33

For the cell where EDUCATION is *grade school*, SEX is *male* and you have 3 subjects observed (cell a), using the little formula above you multiply row total by column total divided by total N

(15 x 9)/33 = 4.1 for cell a

Then for cell b where EDUCATION is *H.S.* and SEX is *male,*

(15 x 9/)33 = 4.1 for cell b

And for the rest,

(15 x 15)33 = 6.8 for cell c

(18 x 9)/33 = 4.1 for cell d

(18 x 9)/33 = 4.9 for cell e

(18 x 15)/33 = 8.2 for cell f

Now putting both **O** and **E** with the expected in parentheses for each respective cell we have:

SEX	EDUC. grade school	H.S.	college	Total
male	3 (4.1)	0 (4.1)	12 (6.8)	15
female	6 (4.9)	6 (4.9)	3 (8.1)	18
Total	9	9	15	33

Now we have all the needed values and can simply go ahead with the calculations. There is no necessary order as long as you do all the cells. So applying the simple algebra to each cell we get:

O	E	(O-E)	$(O-E)^2$
3	4.1	-.1.1	1.21
0	4.1	- 4.1	16.8
12	6.8	5.2	27.0
6	4.9	1.1	1.21
9	4.9	4.1	`6.8
3	8.1	-5.1	26.0
		Sum=	79.0

Table 12.2, *Chi-Square Calculation*

That is the sum for each cell yields the value of chi-square for that table. So X^2 is **79.0**. The value of chi square is always positive.[25]

Next we calculate the degrees of freedom that we will use later as you shall see. The degrees of freedom as used in **t**, **F**, r etc. are determined by N. But in chi-square df is determined by the number of cells. The formula is:

$$df = (r-1)(c-1); \text{(rows-1) times (columns-1)}.$$

Formula 11.2. *Degrees of Freedom for X^2*

Chi-square tests to see whether what you have observed (**O**) is significantly different from what you would expect (**E**) if the null were true. Now we can go ahead and set up the inferential test steps for chi-square.

1. Set alpha at **05,** the usual rejection region for social science.

2. Set the alternate:

$$\text{a. } H_1: X^2 \neq 0$$

3. Set the null, which is 'nothing' going on:

---

25) Although the whole rejection region is on one side it is considered a two-tailed test anyway.

$$H_0: X^2 = 0$$

4. Do the calculations to get the final test statistic, chi-square.(X^2)

This may be done using the method above or, it may be simply the total **N** evenly distributed across the cells (but only when the rows and columns are all equa)l. As a result it is possible

to have a fraction of a whole number as the expected. In other cases (as in our first example) the expected is adjusted proportionately for the relative numbers in each independent variable category (such as 20 males and 14 females figure).

5. Calculate **df.**

So for our 3 x 2 table **df** = (3-1) (2-1) = **2** degrees of freedom.

6. Go to Table E (partially reproduced below) and in this case cross **2 df** (fat left column for degrees of freedom) with the column for **.05**. (highlighted, this is your chosen alpha). The minimum value of your calculated chi-square that you need to reject the null at .05 with 9 cells (**2 df**) is **5.991**. Only if your calculated chi-square is at least that big can you reject the null.

Table E
Chi-Square Probabilities

df	0.995	0.99	0.975	0.95	0.90	0.10	0.05	0.025	0.01	0.005
1	---	---	0.001	0.004	0.016	2.706	**3.841**	5.024	6.635	7.879
2	0.010	0.020	0.051	0.103	0.211	4.605	**5.991**	7.378	9.210	10.597
3	0.072	0.115	0.216	0.352	0.584	6.251	7.815	9.348	11.345	12.838
4	0.207	0.297	0.484	0.711	1.064	7.779	9.488	11.143	13.277	14.860

Table 11.5 *Partial Reproduction of Table E , Chi Square Probabilities*

Since our calculated X^2 is **79.0** which is clearly above **5.991** from Table E, we can easily reject the null that EDUCATION LEVEL does not depend on SEX with a 5% of being wrong when we reject this null.

While test statistics, *r, t, F* etc. are distributed normally, X^2 values are skewed positively. The 5% rejection region is all on the right side (one-sided) yet the test is in fact two-tailed (because it is non-directional). As an example, Figure 10.0 is a graphic of the distribution of chi-square when degrees of freedom are 2. Chi-square on the one hand may be used to test whether some nominal outcome,

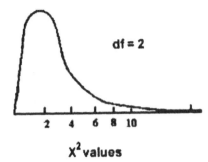

Figure 10.01 X^2 *distribution when df=2*

say SUCCESSFULLY PLACED (1 = successful, 0 = unsuccessful i.e., dichotomous nominal) depends on MARITAL STATUS (1= married 0= not married, also dichotomous nominal). Or perhaps whether EDUCATION (1= completes high school 0= not, dichotomous nominal) depends on RACE (multicategorical).

But chi-square, the identical formula, may also be interpreted as a measure of association. So like the Spearman **r,** you may correlate two ordinal variables, sibling position (three categories) and degree of agreement with a statement (agree, not sure, disagree). For 2 x 2 tables you simply interpret X^2 as the correlation. For larger tables you use phi that is presented later.

Let's do an even easier one. This time we'll use a 2 x 2 table[26]. And we'll do it by hand instead of using SPSS. Once you have your data collected you again restate your hypothesis. In this case our hypothesis is that whether a teen is ARRESTED (1 = yes) or not (0 = not) depends on whether they come from a one-parent (1) or two-parent FAMILY. Our observed data are as follows:

| | ARRESTED | | |
	yes	no	total
	1-parent	**28 a**	**10 b**	38
FAMILY	2-parent	**10 c**	**28 d**	38
	total	38	38	72

Table 11.5 *Observed Data*

Now we go ahead and calculate the expected. Note that when you have a homogenous pattern with equal totals along the margins such as N=38 then you will have the same expected value for each of the cells. You only have to calculate the expected once in that case.

row total X column total = expected for a cell and All cells / total N

---

26) There is such a thing as a table with only one row and two columns but it's use is a bit beyond this text. Such a format can be set up as a binomial probability or as a X2 with an I-R variable as the IV.

For the cell a where ARRESTED is dependent on 1-PARENT you have 28 subjects observed, using the little formula above you multiply row total by column total divided by total N .

$$(38 \times 38)/72 = 4.1 \text{ for cell a and All cells}$$

Now we have all the needed values and can simply go ahead with the calculations. There is no necessary order as long as you do all the cells. So applying the simple algebra to each cell we get:

O	E	(O-E)	(O-E)^2
3	4.1	-.1.1	1.21
0	4.1	- 4.1	16.8
12	6.8	5.2	27.0
6	4.9	1.1	1.21
9	4.9	4.1	`6.8
3	8.1	-5.1	26.0

$$X^2 = 79 \text{ (df 1)}$$

So X^2 is **79.0**. The value of chi square is always positive.[27] Next we calculate the degrees of freedom.

$$\textbf{df} = \textbf{(r-1) (c-1)} = \textbf{(rows-1) x (columns-1).}$$
$$\textbf{df} = \textbf{(r-1) (c-1)} = \textbf{(2-1) x (2-1).}$$
$$\textbf{df} = \textbf{(r-1) (c-1)} = \textbf{1 x 1 = 1}$$

Then using our calculated chi-square of **79 (df 1)** we look it up in the table. Go ahead. You'll see that value of **79** is far above the minimum value of **3.84** needed to reject the null at **p = .05**. So we have significant outcome. That is we may say that being arrested or not does depend on whether the t comes form a one or two-parent family with at most a 5% chance of error in rejecting the null.

Fisher's Exact

Chi-square is sensitive to sample size. If your sample is too large or too small you something else. The Fisher's Exact test is often used for a 2 x 2 table when your sample siz (defined as 80% of the expected frequencies being less than 5). We don't need to go into the here just know that Fisher's exact is available as a chi-square option in statistical software.

Phi

It is unusual to use an inferential test as either a test for group independence c association but this is the case for Chi-Square. You may interpret the standard p-value f

27) Although the whole rejection region is on one side it is still considered to be a two-tailed test any

table as a test for group independence and in the case of a 2 x 2 table you may interpret the p-value associated with chi-square as a measure of association too. Either-or. However if your table is greater than 2 x 2 and you want to detect significant association of the nominal (and sometimes ordinal) variables then you must use *phi*, short for Cramer's phi. The calculation for phi which is reported in the default data output is:

$$\phi = \sqrt{\frac{X^2}{N(k-1)}}$$

Formula 11.3 Phi

Phi is given in terms of a correlation (-1.00 to +1.00) and is interpreted the same way as a Pearson r. Of course it has an associated p-value just as any correlation does.

Summary

1. Chi-square is a non-parametric test, i.e., for nominal or ordinal data.
2. Chi-square can be used as a measure of association but more often is used as a test for group independence.
3. The degrees of freedom are determined by the number of cells, not the number of subjects as in parametric statistics.
4. Ch-square is sensitive to sample size so large sample sizes don't work well. Small sample sizes call for a different non-parametric test, *Fisher's exact*.
5. When determining an association using chi-square the reported output is given in terms of *phi*.

Chapter Eleven: Chi Square - Practice Problems

1. Which test would you use for the following expected values?

a.

(8)	(12)
(20)	(25)

b.

(8)	(2)
(0)	(3)

2. When you want to correlate two nominal level variables you use the _____ test.

3. Calculate chi-square and degrees-of-freedom for the following data. Expected values are in parentheses. Use Table D to state whether the X^2 value is significant at .05.

12 (10)	11 (10)	7 (10)
17 (10)	10 (10)	3 (10)
37 (20)	10 (20)	13 (20)

Chapter Twelve
Single System Analysis

Social work is a practice profession as opposed to a discipline or major such as psychology or sociology. You may have know that the Council on Social Work Education has urged instructors of social work research to include *single system design* in their course content. This is *research on the outcome of an intervention with a single client or family or agency.* That is N = 1. But there are some problems. Remember that all of our analyses so far have used a sample drawn from a population either by convenience or if possible by mathematical probability. We have seen that the bare minimum needed for most analyses is usually ten subjects, preferably thirty or forty. So what do you do when you have only one subject, when N = 1? To answer this, look at the **ABA** diagram below (Figure 11.01). In this case **A** is a period of observation of a targeted behavior and **B** is an intervention period.

This is an ABA design for an individual. We'll call him James. The client is chronically mentally ill and has not been taking his medication for schizophrenia. As a result he uses the local hospital as a revolving door. James' noncompliance is a cost to the hospital and his resulting decompensation also keeps him from maintaining a more supportive routine or social system. Pretend you have James as one of your clients and decide to reinforce medication-compliant behavior by rewarding him with free time on the days when he takes his medication as prescribed. The variable *compliance days* represents the number of days each month that he is compliant with his medication schedule.

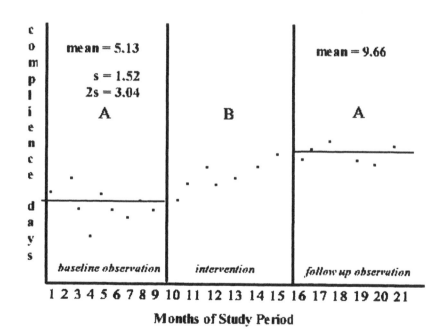

Figure 11.01 *Two Standard Deviation Model*

In the two standard deviation method you calculate the mean and standard deviation from the frequency of days compliant for this single subject during the first baseline period. Normally you would have a group of subjects and calculate the mean number of compliance days per week or per month for the whole group. But again N = 1. You can assume for simplicity's sake that the standard deviations are similar for both the baseline and follow-up periods. So here are the steps involved:

1. Calculate the mean and standard deviation for the baseline period.
2. Add two standard deviations to the mean in the desired direction.
3. Calculate the mean for the follow-up period.
4. Compare the two means to see if the mean for the follow-up period is more or less equal to or greater than the mean of the baseline period plus two standard deviations.

In the example the mean for the follow-up period (9.66 average days compliant per month) is greater than the mean for the baseline observation period plus two standard deviations for the baseline period (5.13 + 3.04 = 8.17). So we can assume that our intervention works.

There are a variety of such designs, AB, ABAB, multiple concurrent designs for different outcomes over the same period etc. In our case we wanted to improve compliance; in another case you might want to reduce some behavior. Your research text should explain the various designs and what is capable of being measured.

Besides the two standard deviation method you might use the *celeration line method*. The celeration line method is a visual inspection where you simply *examine the trend or direction of the slope*. In other words does the slope of direction of change for the observations change during or after the intervention? It should be noted that there are with the reliability and validity of SS design. These issues should be covered in your research.

Single subject designs involve repeated, systematic measurement of a dependent variable before, during, and after the manipulation of an independent variable. Usually, the dependent variable is some characteristic of an individual human being and the independent variable involves the application of some intervention.

There is no experimental control in a traditional case study. Group studies rely upon equivalent groups for experimental control. Single subject studies use individuals as their own controls, comparing no intervention and intervention time periods. Replication across subjects is also used in single subject research to enhance control.

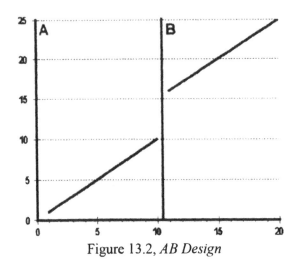

Figure 13.2, *AB Design*

Summary

1. Single subject design can be useful in outcome measurement with an individual client, family or group.
2. It is not technically scientific. There is no target sample or population
3. The advantage is that you can customize interventions. This is difficult to do with samples since people may be very different along a single variable.

Chapter Twelve – Practice Problems

1. Think of two situations that could use single-subject design. For each state:
2. The design, AB, ABA, celebrated line, two-standard deviation method etc.
3. How long will your baseline be?
4. How long with the intervention period be?
5. How long will the follow-up be?
6. What will you assess?
7. How could the outcome be useful if positive, if negative?

Table A
Table of Random Numbers

39634	62349	74088	65564	16379	19713	39153	69459	17986	24537
14595	35050	40469	27478	44526	67331	93365	54526	22356	93208
30734	71571	83722	79712	25775	65178	07763	82928	31131	30196
64628	89126	91254	24090	25752	03091	39411	73146	06089	15630
42831	95113	43511	42082	15140	34733	68076	18292	69486	80468
80583	70361	41047	26792	78466	03395	17635	09697	82447	31405
00209	90404	99457	72570	42194	49043	24330	14939	09865	45906
05409	20830	01911	60767	55248	79253	12317	84120	77772	50103
95836	22530	91785	80210	34361	52228	33869	94332	83868	61672
65358	70469	87149	89509	72176	18103	55169	79954	72002	20582
72249	04037	36192	40221	14918	53437	60571	40995	55006	10694
41692	40581	93050	48734	34652	41577	04631	49184	39295	81776
61885	50796	96822	82002	07973	52925	75467	86013	98072	91942
48917	48129	48624	48248	91465	54898	61220	18721	67387	66575
88378	84299	12193	03785	49314	39761	99132	28775	45276	91816
77800	25734	09801	92087	02955	12872	89848	48579	06028	13827
24028	03405	01178	06316	81916	40170	53665	87202	88638	47121
86558	84750	43994	01760	96205	27937	45416	71964	52261	30781
58697	31973	06303	94202	62287	56164	79157	98375	24558	99241
38449	46438	91579	01907	72146	05764	22400	94490	49833	09258

Table B
Areas Under the Curve by Z-score

z	x.x0	x.x1	x.x2	x.x3	x.x4	x.x5	x.x6	x.x7	x.x8	x.x9
0.0x	.0000	.0040	.0080	.0120	.0160	.0199	.0239	.0279	.0319	.0359
0.1x	.0398	.0438	.0478	.0517	.0557	.0596	.0636	.0675	.0714	.0753
0.2x	.0793	.0832	.0871	.0910	.0948	.0987	.1026	.1064	.1103	.1141
0.3x	.1179	.1217	.1255	.1293	.1331	.1368	.1406	.1443	.1480	.1517
0.4x	.1554	.1591	.1628	.1664	.1700	.1736	.1772	.1808	.1844	.1879
0.5x	.1915	.1950	.1985	.2019	.2054	.2088	.2123	.2157	.2190	.2224
0.6x	.2257	.2291	.2324	.2357	.2389	.2422	.2454	.2486	.2517	.2549
0.7x	.2580	.2611	.2642	.2673	.2704	.2734	.2764	.2794	.2823	.2852
0.8x	.2881	.2910	.2939	.2967	.2995	.3023	.3051	.3078	.3106	.3133
0.9x	.3159	.3186	.3212	.3238	.3264	.3289	.3315	.3340	.3365	.3389
1.0x	**.3413**	.3438	.3461	.3485	.3508	.3531	.3554	.3577	.3599	.3621
1.1x	.3643	.3665	.3686	.3708	.3729	.3749	.3770	.3790	.3810	.3830
1.2x	.3849	.3869	.3888	.3907	.3925	.3944	.3962	.3980	.3997	.4015
1.3x	.4032	.4049	.4066	.4082	.4099	.4115	.4131	.4147	.4162	.4177
1.4x	.4192	.4207	.4222	.4236	.4251	.4265	.4279	.4292	.4306	.4319
1.5x	.4332	.4345	.4357	.4370	.4382	.4394	.4406	.4418	.4429	.4441
1.6x	.4452	.4463	.4474	.4484	.4495	.4505	.4515	.4525	.4535	.4545
1.7x	.4554	.4564	.4573	.4582	.4591	.4599	.4608	.4616	.4625	.4633
1.8x	.4641	.4649	.4656	.4664	.4671	.4678	.4686	.4693	.4699	.4706
1.9x	.4713	.4719	.4726	.4732	.4738	.4744	**.4750**	.4756	.4761	.4767
2.0x	.4772	.4778	.4783	.4788	.4793	.4798	.4803	.4808	.4812	.4817
...	...	...	...	...	...	...	...	...	...	...
3.0x	**.4987**	.4987	.4987	.4988	.4988	.4989	.4989	.4989	.4990	.4990

Table C
Pearson Product-Moment Correlation Coefficients

Level of Significance Two-Tailed Test	.10	.05	.02	.01
df (n-2)	r	r	r	r
1	.988	.997	.9995	.9999
2	.900	.950	.980	.990
3	.805	.878	.934	.959
4	.729	.811	.882	.917
5	.669	.754	.833	.874
6	.622	.707	.789	.834
7	.582	.666	.750	.798
8	.549	.632	.716	.765
9	.521	.602	.685	.735
10	.497	.576	.658	.708
11	.476	.553	.634	.684
12	.458	.532	.612	.661
13	.441	.514	.592	.641
14	.426	.497	.574	.623
15	.412	.482	.558	.606
16	.400	.468	.542	.590
17	.389	.456	.528	.575
18	.378	.444	.516	.561
19	.369	.433	.503	.549
20	.360	.423	.492	.537
21	.352	.413	.482	.526

22	.344	.404	.472	.515
23	.337	.396	.462	.505
24	.330	.388	.453	.496
25	.323	.381	.445	.487
26	.317	.374	.437	.479
27	.311	.367	.430	.471
28	.306	.361	.423	.463
29	.301	.355	.416	.456
30	.296	.349	.409	.449
35	.275	.325	.381	.418
40	.257	.304	.358	.393
45	.243	.288	.338	.372
50	.231	.273	.322	.354
60	.211	.250	.295	.325
70	.195	.232	.274	.303
80	.183	.217	.256	.283
90	.173	.205	.242	.267
100	.164	.195	.230	.254

Table D
t-test Probabilities

Levels of significance for <u>one</u> tailed tests				
df	0.05	0.025	0.01	0.005
Levels of significance for <u>two</u> tailed tests				
df	0.010	**0.05**	0.02	0.01
5	2.015	2.571	3.365	4.032
6	1.943	2.447	3.143	3.707
7	1.895	2.365	2.998	3.499
8	1.860	2.306	2.896	3.355
9	1.833	2.262	2.821	3.250
10	1.812	2.228	2.764	3.169
11	1.796	2.201	2.718	3.106
12	1.782	2.179	2.681	3.055
13	1.771	2.160	2.650	3.012
14	1.761	2.145	2.624	2.977
15	1.753	2.131	2.602	2.947
16	1.746	2.120	2.583	2.921
17	1.740	2.110	2.567	2.898
18	1.734	2.101	2.552	2.878
19	1.729	2.093	2.539	2.861

Tables

Table E
Value of F = 0.05
(confidence = 95%) for small samples

Vertical Axis = <u>Denominator</u> Degrees of Freedom for (MS_b)
Horizontal Axis = <u>Numerator</u> Degrees of Freedom for (MS_w)

df	1	2	3	4	5	9	14	19	24
1	161	200	216	224	230	240	245	248	249
2	18.5	19.0	19.2	19.2	19.3	19.4	19.4	19.4	19.4
3	10.1	9.55	9.28	9.12	9.01	8.81	8.71	8.67	8.64
4	7.71	6.94	6.59	6.39	6.26	6.00	5.87	5.81	5.77
5	6.61	5.79	5.41	5.19	5.05	4.77	4.64	4.57	4.53
9	5.12	4.26	3.86	3.63	3.48	3.18	3.02	2.95	2.90
14	4.60	3.74	3.34	3.11	2.96	2.64	2.48	2.40	2.35
19	4.38	3.52	3.13	2.90	2.74	2.42	2.26	2.17	2.11
24	4.26	3.40	3.01	2.78	2.62	2.30	2.13	2.04	1.98